Lecture Notes in Computer Science 5941

Commenced Publication in 1973
Founding and Former Series Editors:
Gerhard Goos, Juris Hartmanis, and Jan v

Editorial Board

Juraj Hromkovič
Richard Královič
Jan Vahrenhold (Eds.)

Teaching Fundamental Concepts of Informatics

4th International Conference on Informatics in Secondary
Schools - Evolution and Perspectives, ISSEP 2010
Zurich, Switzerland, January 13-15, 2010
Proceedings

 Springer

Volume Editors

Juraj Hromkovič
Richard Královič
ETH Zürich
Informationstechnologie und Ausbildung
CAB F 16, F 13.1
Universitätstrasse 6
8092 Zürich, Switzerland
E-mail: {juraj.hromkovic, richard.kralovic}@inf.ethz.ch

Jan Vahrenhold
Technische Universität Dortmund
Foundations of Computer Science and Computer Science Education Group
Chair of Algorithm Engineering
Faculty of Computer Science
Otto-Hahn-Str. 14
44227 Dortmund, Germany
E-mail: jan.vahrenhold@cs.tu-dortmund.de

Library of Congress Control Number: 2009941784

CR Subject Classification (1998): K.3, J.1, K.8, H.5.2, D.1, D.3

LNCS Sublibrary: SL 1 – Theoretical Computer Science and General Issues

ISSN   0302-9743
ISBN-10   3-642-11375-3 Springer Berlin Heidelberg New York
ISBN-13   978-3-642-11375-8 Springer Berlin Heidelberg New York

springer.com

© Springer-Verlag Berlin Heidelberg 2010
Printed in Germany

Typesetting: Camera-ready by author, data conversion by Scientific Publishing Services, Chennai, India
Printed on acid-free paper   SPIN: 12830671   06/3180   5 4 3 2 1 0

Preface

The International Conference on Informatics in Secondary Schools: Evolution and Perspective (ISSEP) is an emerging forum for researchers and practitioners in the area of computer science education with a focus on secondary schools.

The ISSEP series started in 2005 in Klagenfurt, and continued in 2006 in Vilnius, and in 2008 in Toruń. The 4th ISSEP took part in Zurich. This volume presents 4 of the 5 invited talks and 14 regular contributions chosen from 32 submissions to ISSEP 2010.

The ISSEP conference series is devoted to all aspects of computer science teaching. In the preface of the proceedings of ISSEP 2006, Roland Mittermeir wrote: "ISSEP aims at educating 'informatics proper' by showing the beauty of the discipline, hoping to create interest in a later professional career in computing, and it will give answers different from the opinion of those who used to familiarize pupils with the basics of ICT in order to achieve computer literacy for the young generation." This is an important message at this time, when several countries have reduced teaching informatics to educating about current software packages that change from year to year. The goal of ISSEP is to support teaching of the basic concepts and methods of informatics, thereby making it a subject in secondary schools that is comparable in depth and requirements with mathematics or natural sciences. As we tried to present in our book "Algorithmic Adventures. From Knowledge to Magic," we aim at teaching informatics as a challenging scientific discipline, full of puzzles, challenges, magic and surprising discoveries. Additionally, this way of teaching informatics is also a chance to import the concept of engineering to schools, by merging the mathematical analytic way of thinking with the constructive work of engineers in the education of one subject.

To underline informatics as well as informatics didactics as scientific disciplines, ISSEP 2010 had two special tracks. The track "Contributions of Competitions to Informatics Education" was based on the fact that taking part in different kinds of competitions provides a valuable contribution to knowledge acquirement and supports the development of problem-solving skills in a creative way. Organizing a competition includes addressing the following two questions:

- Which kinds of competitions are especially well suited for achieving which goals?
- How should one create and choose tasks and rules for such competitions?
- What are the achievements of the competition participants, in particular in relation to their training process?
- What is the influence of competitions on the educational processes in secondary education?

The starting point to this track was provided by the invited talk "Sustaining Informatics Education by Contests" by Valentina Dagienė.

The second track, "Empirical Research," pointed out that the community of computer science didactics has to strengthen its effort in empirical research in order to be as serious as the didactics of mathematics and physics are. The main questions posed were:

- What is "good empirical research?"
- Which rules should be followed to produce "good" empirical results?
- Which criteria can be applied to recognize "good" empirical results?
- What are the pitfalls of interpreting empirical results?

To make ISSEP 2010 attractive due to high-quality contributions, we increased the number of invited speakers to five. In addition to Valentina Dagienė (Vilnius), we invited the internationally leading experts David Ginat (Tel Aviv University), David Gries (Cornell University), Allen B. Tucker (Bowdoin College), and Amiram Yehudai (Tel Aviv University) to give talks about different aspects of computer science education.

I would like to express my deepest thanks to all members of the Program Committee for serving and thus contributing to the high standard of the ISSEP series among the conferences devoted to computer science education.

November 2009   Juraj Hromkovič

Conference Organization

Program Chairs

Juraj Hromkovič   ETH Zürich
Jan Vahrenhold   TU Dortmund

Program Committee

Peter Antonitsch	University of Klagenfurt
Owen L. Astrachan	Duke University
Ralph-Johan Back	Abo Akademi University
Harry Buhrman	CWI & University of Amsterdam
Valentina Dagienė	Institute of Mathematics and Informatics, Vilnius
Judith Gal-Ezer	The Open University of Israel
David Ginat	Tel Aviv University
Peter Hubwieser	TU München
Ivan Kalaš	University of Bratislava
Peter Micheuz	University of Klagenfurt
Roland Mittermeir	University of Klagenfurt
Wolfgang Pohl	Bundeswettbewerb Informatik
Ulrik Schroeder	RWTH Aachen
Jarkko Suhonen	University of Joensuu
Maciej M. Sysło	UMK Torun, University of Wroclaw
Tom Verhoeff	TU Eindhoven
Michal Winczer	UK Bratislava

External Reviewers

Brauner, Philipp
Leonhardt, Thiemo

Local Organization

Juraj Hromkovič (Chair)	ETH Zürich
Hans-Joachim Böckenhauer	ETH Zürich
Herbert Bruderer	ETH Zürich
Petra Hieber	ETH Zürich
Blanca Höhn	ETH Zürich
Lucia Keller	ETH Zürich
Dennis Komm	ETH Zürich
Richard Královič	ETH Zürich

Jan Lichtensteiger	ETH Zürich
Ueli Marty	ETH Zürich
Tobias Mömke	ETH Zürich
Giovanni Serafini	ETH Zürich
Andreas Sprock	ETH Zürich
Björn Steffen	ETH Zürich
Monika Steinová	ETH Zürich

Table of Contents

Sustaining Informatics Education by Contests

Valentina Dagienė

Institute of Mathematics and Informatics
Akademijos str. 4, LT-08663 Vilnius, Lithuania
dagiene@ktl.mii.lt

Abstract. Three decades ago high school computing was highly consistent with academic and professional world. This consistency was destroyed when school curricula began to emphasize information and communication technology skills at the expense of computer science. Recently many countries began to think how to re-establish informatics education in schools and how to attract pupils to choose optional modules related to computer science. Although informatics is not taught as a discipline in many countries, pupils are invited to participate in different contests on informatics organized all over the world. When pupils get interested in programming contests, they are looking for training and gain some informatics education. Contests are exceptionally valuable for motivating and involving pupils in computer science. The current paper discusses the contests and olympiads in informatics arranged internationally and continuously. The main attention is paid to the model of International Olympiad in Informatics and International Contest on Informatics and Computer Fluency (named Bebras in Lithuanian, or Beaver in English).

Keywords: Teaching informatics, computer science education, teaching programming, olympiad in informatics, contest on informatics.

1 What? Why? How? – Questions That Should Be Reconsidered in Informatics Education

In one of the fundamental papers on teaching informatics, Juraj Hromkovic asked: What is informatics? What is computers science? Why teach computer science? What to teach and how to teach it? [1] These are core questions to everybody who has been thinking on bringing informatics to the school level.

Significant changes in society do not begin on one particular day or even in a particular year. Changes come slowly, especially in education. Teachers, policy makers and researchers should work continuously for decades in order to gain significant results on pupils' achievements in informatics.

The education achievements that were obtained in the eighties and the nineties of the 20th century might be explained by the implementation of computers and information technologies (IT) in schools and by forming of their impact to general education. In Europe and world wide, countries were tackling the problem in different ways: richer countries were buying computers on a mass scale and supplying their schools with educational software. They were also arranging training courses for

J. Hromkovič, R. Královič, and J. Vahrenhold (Eds.): ISSEP 2010, LNCS 5941, pp. 1–12, 2010.

teachers. Thus, they were using computers and modern technology wherever they could. The majority of other countries, on the other hand, were trying to develop theoretical well-grounded models of informatics and IT education, compose curricula, syllabi, tutorials, textbooks, and arrange trainings, i.e. at implementation to all students with moderate investment in equipment.

In the beginning of the 1980s, the informatics education, although in a different range, was established in the majority of schools of some countries, e.g. Austria [2, 3], Germany, Lithuania [4], Russia. The German computer scientist Klaus Haefner warned that education should be adopted quickly to avoid the risk of misqualifying people [5]. The human brain would be challenged by the growth of information technology and would be subject to competition of information processing systems. To overcome the crises mentioned in time, Haefner recommended bringing informatics into the classes and developing new curricula with information technology behind.

One of the early Russian pioneers in the field of theoretical and systems programming, a founder of the Siberian School of Computer Science Andrei Ershov, has declared a slogan: "Programming is the second literacy" [6]. It has become a popular metaphor, which has been widely used around the world. Politicians and educators in the industrialized countries, proclaimed "computer literacy' as an essential part of education and they demanded the integration of new technologies into the school curriculum.

Teaching informatics started with programming. Sometimes it was interpreted that machines at that time were miserable and that programming was the exceptional possibility to manage them. However, the goal of teaching programming is problem solving transfer, i.e., users are expected to be able to apply what they have learned to solving problems that they have not been taught [7]. Furthermore, programming is the best way to build a language for instructing (communicating) a machine. According to Hromkovic "We have to teach programming as a skill to describe possibly complex behaviors by a sequence of clear, simple instructions" [1, p. 33]. Later, Avi Cohen and Bruria Haberman went further and declared computer science as a language of technology [8].

A significant role in designing methodology for teaching programming has been played by the scientists of Lithuania. Already in 1978–1979, a students' education in programming by using postal services was drafted. After accomplishment of certain experiments, the Young Programmer's School by Correspondence was established in 1981 [9]. This is one of the oldest schools for teaching programming and it continues to function nowadays. The activity of the Young Programmer's School in distance learning was one of the first examples concerning informatics and had a strong impact on many phenomena related to informatics' teaching, such as accomplishment of the UNESCO initiated project "Distance learning of informatics (programming)" in 1992–1993 [10] and development of the Contests and Olympiads in Informatics [11].

In the recent years, enrolments in the undergraduate programs of computer science have been dropping. There are many factors that have contributed to the decline in student interest, some of which relate to the lack of understanding the essence of computing at school [12]. These after-effects are very closely connected with what has been done in many western countries: computer science was exchanged for information and communication technology in schools. "… we, as computer scientists, are also responsible for this big misunderstanding…", declared Hromkovic [1, p. 25].

Nowadays more and more countries have been reconsidering the role of informatics in general education, e.g. France is discussing the curricula for teaching informatics in secondary schools, and Slovakia is developing new courses for teacher training in informatics.

Bringing informatics to schools through curriculum in a formal track is quite important, however it is necessary to support the informal ways of introducing pupils to informatics. The most famous informal way to introduce informatics are contests and olympiads on programming [13, 14].

Contests make teaching of programming more attractive for students. Furthermore, computer programming is one of the appropriate and effective ways to develop problem solving skills for computer science learners [15]. During contests students meet their peers from all around the country (or countries), make friends, and wait for the next competition ready to show their abilities which have improved since the last contest.

2   Contests on Programming for General Education

Developing abilities to master modern technologies and skills for solving problems is among the most important capabilities of an educated future citizen of an information society and it can be straight connected with informatics education. Problem solving by means of programming does not lose its importance in a contemporary school equipped with modern information technologies and it will remain as a very important part of understanding the information processing and running computer. Programming, with emphasis on algorithms, remains the core of several worldwide contests, e.g. International Olympiad in Informatics (IOI) and the USA Computing Olympiad (USACO). The USACO holds six internet-based contests each year and has several difficulty divisions [16].

In developing teaching of programming, we recommend considering the attractiveness of instructional methods and consolidation of pupils' motivation. The following aspects should be taken into account:

- For school students, practical activities are much more interesting and attractive than academic studies.
- Elements of contests and competition stimulate the learning process.

More time should be dedicated to the motivation, aims, connection between practice and theoretical concepts, and especially to the internal context of the presented theory.

Programming is an activity composed of several components: comprehension of the problem, choosing algorithm, encoding it, debugging, testing, and optimizing [17]. Since many of the skills required for successful programming are similar to those required for effective problem solving, computer programming and particularly choosing one of several possible solutions and later debugging in a short period of time, provides a fertile field for developing and practicing problem solving skills in an environment that is engaging for young students [18].

When students begin learning basics of programming, they soon try to find a place where they can demonstrate their skills, their projects, share their interests or compare themselves with others. This might explain the reasons why many students, soon after they have started learning programming, choose one of the areas where they are able to demonstrate their work immediately, e.g. creation of web pages, or computer graphics. For some areas, e.g. developing algorithms, it is not easy to find practical demonstration. The most powerful means which endorse students' motivation are competitions or contests.

There, the pupils meet their peers from all over the country and form other countries; they make friendships, wait for the next contest ready to show their abilities which have improved since last contest. In the programming contests, pupils use and develop, at the same time, their problem solving skills. Furthermore, pupils especially gifted can be challenged by problems that cannot be solved by applying learned mechanisms, but that require special talent, mental abilities, and probably extraordinary effort, too.

Pupils like to be involved in competition, they like to compete [19]. In education, it is important to find right place for competition: these can be contests or challenges. In a contest, the main interest is the quality of the individual performance; contestants are confronted with problems, not with each other. Contests are extracurricular activities that allow students to acquire their knowledge and, understanding from the classroom, apply it within a competitive environment. These types of activities provide ways of challenging students in creative and innovative ways.

There have been many academic competitions and contests in computer science throughout allover the world. Most of them are programming contests with focus on algorithmic problem solving. There are several contests covering other scientific areas, most prominent examples are contests in robotics: Robocup Junior and First League. There are mixed contests that cover different areas, for example, the American Computer Science League (ACSL). The contests of the ACSL mostly consist of a short answer test and a programming problem. A short answer test contains five questions from categories like number systems, logic, Lisp, data structures, graph theory, digital electronics and WDTPD (What Does This Program Do). Typically answers are very short. The programming problem is solved by submitting a program source code within 72 hours. Framework of classification on computer science contests for secondary school students is provided by Wolfgang Pohl in [20].

There are two main paradigms for implementing contests: from an international level to the local one (top-down strategy), and vice versa, from local activities to an international promotion (bottom-up strategy). The first paradigm is a challenge to find some suitable international contests, analyze, train students, and join them after intensive work. The second paradigm stresses an opportunity to establish the local contest and attempt to develop it to an international level. The IOI is a contest referred to the first competition paradigm while the Bebras International Contest on Informatics and Computer Fluency [21] belongs to the second paradigm.

2.1 International Olympiads in Informatics

The IOI is one of the five international science Olympiads initiated by UNESCO in 1987. It is an annual international informatics competition for individual contestants from many countries around the world, accompanied by social and cultural programs [22].

These competitions focus on informatics problems of algorithmic nature. In the scope of IOI the concept *Informatics* means a domain that is also known as computer science, computing science and information technology.

Yet, the high-level goal of the IOI is to promote computer science among the youth, and to stimulate their interest in programming and algorithms. The contest brings exceptionally gifted pupils from various countries together and renders them an opportunity to share scientific and cultural experiences. Thus, one of the main objectives in each country is to discover, encourage and train exceptionally talented young people in computer science.

The IOI is managed by the General Assembly, which is a temporary, short-term committee composed of the leaders of all the participating countries and by two long standing committees. The International (Steering) Committee consists of representatives of the past, present, and future IOI's as well as several elected representatives. Its task is to retain the continuity of the IOI by finding future host countries. The second committee is the IOI Scientific Committee, the task of which is to ensure continuity and quality control of the IOI competitions [23].

The IOI is organized in and by one of the participating countries. Each participating country typically sends a delegation of four students accompanied by two leaders. Students are usually selected in the national olympiads in informatics or programming contests. Each of the two competition days lasts for five hours with 3 or 4 tasks to be solved.

The students compete individually and try to maximize their score by solving a set of problems. The IOI contestants are required to express their algorithms in one of the allowed programming languages (currently Pascal and C/C++) and they must engineer their programs to run flawlessly, because marking is based on automated execution [24].

Organized in 1989 in Pravec, Bulgaria, the IOI celebrated 20 years anniversary again in Bulgaria, this time in Plovdiv. The 101 tasks were presented for students during 20 years. Tom Verhoeff, one of the leading persons in developing tasks for the IOI, analyzed the 20-year history of IOI tasks and summarized task type and difficulty level, and classified them according to concepts involved in their problem and solution domain [25]. Difficulty level is determined on the basis of what percentage of contestants were able to 'fully' (a submission should be scored 90 % or more) solve the task. According to Verhoeff, many of the tasks are too difficult to use 'as is' in regular computer science courses for secondary education [25].

The most significant contribution of the IOI to computer science education can be considered olympics movements in many countries and regions. Only 13 countries participated in the first IOI, whereas already 82 countries were involved in the 21st Olympiad (actually 79 countries with participating teams and 3 countries observers). Almost all these countries organize national contests or olympiads in informatics and train pupils and teachers. Some of these contests were implemented following the IOI model (with some adaptation to national peculiarities), although some countries are concentrated on their own infrastructure of contests. Additionally there are regional olympiads in informatics, e.g. African, Asian, Arabic, Balkan, Baltic, Central European; usually they are organized in the same manner as the IOI.

2.2 The IOI Conference on Olympiads in Informatics

The IOI community consists of about 80 participating countries. We face mainly the same problems: how many of the countries have national olympiads in informatics? Do they have some other contests on programming? How do we pick our students? How do we train them? The IOI presents an ideal forum for discussing these experiences and associated issues. It was decided to establish conferences during IOIs.

The first IOI conference "Olympiads in Informatics" was held in Zagreb, Croatia in 2007 during the first and second competition days. The 17 selected papers discussed the running of and issues facing several national olympiads: Brazilian [26], Canadian [27], Chinese [28], Croatian [29], German [30], Italian [31], Polish [32], Russian [33], etc.

Next IOI conference was organized in Cairo, Egypt. It concentrated on training and task types, and many of the ideas and experiences were drawn from the national olympiads. Tasks are perennial issue for contests, their most visible aspect and, for many contestants, the primary reason for participation. The IOI community strives for quality, variety and suitability. We endeavour to make tasks interesting, understandable and accessible.

Fourteen research papers were published and discussed during the third IOI conference held in Plovdiv, Bulgaria in 2009. The major part of the papers were focussed on training and task types, and some of the ideas and experiences are drawn from the national olympiads.

Tasks are perennial issue for contests, their most visible aspect and, for many contestants, the primary reason for participation. We strive for quality, variety and suitability. If to the contestants, tasks seem to be the main purpose of an olympiad, from an educator's perspective there is often an equal interest in training the contestants. This is not only a question of how we choose the best, or enable to show their true ability. We seek to enthuse them with a passion for the subject.

The IOI conferences are followed by the journal "Olympiads in Informatics" (http://www.mii.lt/olympiads_in_informatics). It is a refereed scholarly journal that provides an international forum for presenting research and development in teaching and learning informatics through competition. Three already published volumes have been closely connected with the IOI conference. Starting from this year, submissions of papers are flexible and there are no requirements to participate in the conference while paper is accepted.

2.3 Regional and National Olympiads in Informatics

The national olympiads exist in a wider community – of course, it is also true for the international olympiads. In order to ensure better preparation for the IOI and to strengthen regional relations, various regional olympiads are organized. While the national olympiads represent informatics teaching traditions of each country, the regional olympiads are usually a mini model of the IOI, allowing the participants to experience what they will come through in the IOI.

We shortly present one of the regional olympiads, the Baltic Olympiad in Informatics (BOI); more detailed view can be found in [34]. The BOI was established on the initiative of three Baltic countries (Estonia, Latvia, and Lithuania) in 1995, and few years later it was opened to all countries around the Baltic Sea. The main goals concentrate on providing the participating students with experience of an international olympiad, encouraging communication and exchange of ideas between the developers of national contests in informatics, as well as assisting delegation leaders in selecting participants for the IOI.

The BOI is a short-term (lasting 3-4 days) and inexpensive event. It can be distinguished for cozy and good neighborly atmosphere, which is highly important when motivating students for self-help.

Even though the BOI is a mini-model of the IOI, it has significant differences from the cultural and learning perspectives. The organization of the scientific part of BOI's is based on mutual trust of the participating countries. The leaders of all the participating countries offer problems for the nearest BOI. At first draft task texts are offered, then the ideas are exchanged via e-mail, discussed, some problems rejected, while other problems are suggested to be modified and later are accepted. Most of the problems are translated to the native languages by the leaders before going to the olympiad. This is a unique possibility for country representatives to gain experience in organizing the scientific part of a small international olympiad as well as to raise their qualifications in algorithms.

The BOI is also a form of learning for its participants. On one hand, they come to the event ready to gain some international experience after participating in the domestic contest. On the other hand, they know that their final destination is the IOI, and they try to learn as much as possible in the BOI. The organizers of BOI's try to follow as close as possible the newest IOI trends in problem types, compilers, platforms, and contest systems. Even though all the tasks are of the algorithmic nature, they represent cultural and methodical differences. Since in the BOI much preparatory work has been done in advance, team leaders can discuss the tasks, possible solutions and technical issues and the BOI can be considered as a pre-arranged international way of learning.

National olympiads have more complex infrastructure. Countries have different educational systems but there are as many similarities as differences. Usually there are several rounds, e.g. school, regional or state round. It is also common to use difficulty levels e.g. junior and senior.

3 International Contest on Informatics and Computer Fluency

Olympiads in Informatics focus mainly on developing algorithms and programming. Programming is an important area of computer science but it is not enough. What else could we suggest to pupils at secondary schools so that they would gain more complete view of computer science? Which topics of computer science are "compulsory" to general education?

We agreed that teaching informatics in secondary or even primary schools has to start with programming. The programming course can be followed by the introduction of some basic concepts of data structures, recursions, procedures, and fundamental methods for designing algorithms.

Hromkovic suggested to start teaching fundamentals with automata theory [1, p. 34]. Using simple mathematics, pupils can learn modular design methodology of hardware systems and some important verification concepts.

These and other computer science concepts can be introduced to pupils again by contest. The contest should be the key to present various topics of computer science in an attractive way [35].

The idea of the Bebras Contest on Informatics and Computer Fluency originated in Lithuania in 2003 (the name Bebras – in English "beaver" – is connected with a hard-working, intelligent, goal seeking and lively animal). It took almost a year to create the tasks and to prepare the technology for implementing that: the first contest started in October 2004 [36, 37]. The Bebras contest started in a coordinated way: run of contests at schools, where solutions may be submitted to some central authorities or some local organizers.

The Bebras contest addresses all lower and upper secondary school dividing pupils into three age groups: Benjamin (age 11-14), Junior (age 15-16), and Senior (upper secondary level). Some countries divide Benjamin's group further in Cadets (age 13-14).

Any contest needs to have first, a challenging set of tasks and second, a grading procedure. Tasks are the most important. The Bebras tasks' developers are seeking to choose interesting tasks (problems) for motivating students to deal with computer science and to think deeper about technology. Some agreements on tasks development criteria have to be settled [38, 39].

In the past few years, the number of the Bebras participants has been notably growing. In 2007, the Bebras contests took place in seven countries, with about 50 000 participants total. In 2008, more than 90 000 students from 10 countries played the game, world-wide [21]. Estonia had the strongest relative participation with 4 039 contestants, whereas Germany had the highest total number of participants, exactly 53 602. Seven further countries are going to run Bebras contests (Bulgaria, Czech Republic, Egypt, Finland, Italy, Israel, and Macedonia).

In [39], the six task topics (types) are discussed: (1) information comprehension, (2) Algorithmic thinking, (3) Using computer systems, (4) Structures, patterns and arrangements, (5) Puzzles, (6) Social, ethical, cultural, international, and legal issues. The descriptions of these task types involve also concepts of informatics although this was not the goal of this classification. It gives anyway a rough idea what kinds of problems and what topics of computer science we have in mind for Bebras contest. In short, Bebras tasks can involve concepts of informatics like algorithms and programs: sequential and concurrent; data structures like heaps, stacks and queues; modelling of states, control flow and data flow; human-computer interaction; graphics; etc.

Task example 1. Beaver Creek (Bebras Contest, 2008; type: algorithmic thinking; level: medium, juniors). There are tracks in Beaver Creek. Since beavers do not go backwards, there are some parallel tracks to give the way. Look at the figures. There can be only one beaver in each cell. In which situation is a total traffic jam *unavoidable*?

To solve this task one has to imagine how the beavers walk from one field to the next, where the beavers do their steps, and whether they do one step after each other or walk in parallel! With this task one can learn that concurrent activities may end up in a deadlock where no further move is possible. This task allows an insight into autonomous agents that act in parallel and may run in a so-called deadlock.

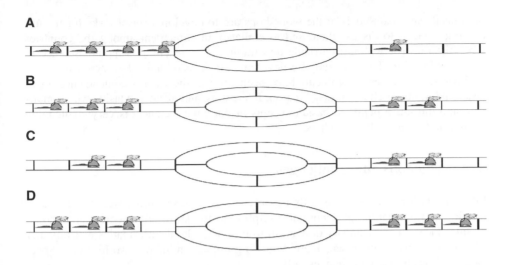

Task example 2. Paperchase. (Bebras Contest, 2007; type: structures, patterns, arrangements; level: hard, seniors). Peter writes on a paper the letters he finds on his way following the arrows. Some of the arrows have no letter. Which one of the following sequences of letters *cannot* be written by Peter on his way from *Start* to *Finish*?

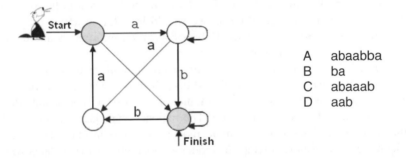

A   abaabba
B   ba
C   abaaab
D   aab

Task example 3. Drawing spiral. Using Logo, a simple programming language, Anna has drawn a rectangular spiral with the help of the following commands:

forward 10 – the turtle moves forward drawing a line of 10 steps (dots) long;
left 90 – the turtle turns left making an angle of 90 degrees.

Which of the following numbers expresses the length of the whole spiral in dots?

A   550
B   170
C   300
D   250

Since 2004, the Bebras contest runs every autumn, usually at the end of October or at the beginning of November. Before the annual contest, the Bebras workshop is organized. The main goals of the workshops are to develop a set of tasks for the upcoming contest, to discuss them and to come to an agreement among the countries with different (or without) curricula and traditions of teaching computer science in general education. The workshop for developing tasks is organized every spring.

Two general types of problems have been used in the Bebras contest: interactive tasks and multiple-choice tasks. Creating interesting and attractive tasks, which motivate and introduce pupils to basic concepts of computer science, is very challenging for researchers as well as teachers.

4 Conclusions and Challenges

Informatics, the science of algorithmic processing, representation, storage and transmission of information, is important discipline in the knowledge society and should be introduced in to secondary or even in to primary school. Informatics or Computer Science is a fascinating research are with a big impact on the real world, full of spectacular ideas and great challenges [40].

There is a lack of recognition for the value of informatics in general education and the mistake of over-focusing on computer driving skills. There is a shortage of the informatics teaching methodology for schools. We have very few reasonable didactical approaches for teaching informatics, and one of them could be introducing informatics through contests.

Contests play an important role as a tool of motivation and inspiration. In order to encourage students to learn computer science, we should look for attractive didactic forms. The olympiad in informatics and the Bebras contest on informatics and computer fluency could serve as useful examples.

Olympiad is a globally recognized way to involve pupils into informatics and a very important motive to improve their programming skills. There is a steady infrastructure of international olympiad in informatics. A community of scientists, teachers, and other professionals in education has been formed too. The regional olympiads are organized following the same principles. Similar olympic movements exist in many countries. Various difficulty levels in national olympiads render a possibility for the students with different experience and knowledge of programming to participate in the event. Even the beginners in programming can acquire motivation to participate and to learn. Olympiads are mainly focused an algorithms design and programming. Actually, tasks are rather difficult for many pupils and require continuous efforts.

The Bebras international contest on informatics and computer fluency is established with an idea that it should fit to each pupil regardless of whether she or he is taught informatics at school or not. The main goals of the Bebras contest are to evoke interest in computer science for everyone, to motivate pupils to understand its fundamentals and to be fluent with the technology, e. g to be able to communicate with a machine. The contest should help children to get interested in informatics and to stimulate thinking about contributions of informatics to science at the very beginning of school.

Let's take olympiads and contests in informatics as a serious didactical approach of computer science education and support them with learning material, tools, teacher training courses, etc. Olympic and contest communities are open for proposals and ideas for collaboration and future developments. Involving pupils in recognition of informatics as a science discipline should be our target, and we should try to achieve it together. Well-organized contests with interesting, playful, exciting problems, and attractive awards will involve pupils into the essence of the computer science world and will help to understand the realities, possibilities, and failings of the technology.

References

1. Hromkovic, J.: Contributing to General Education by Teaching Informatics. In: Mittermeir, R.T. (ed.) ISSEP 2006. LNCS, vol. 4226, pp. 25–37. Springer, Heidelberg (2006)
2. Micheuz, P.: 20 Years of Computers and Informatics in Austria's Secondary Academic Schools. In: Mittermeir, R.T. (ed.) ISSEP 2005. LNCS, vol. 3422, pp. 20–31. Springer, Heidelberg (2005)
3. Reiter, A.: Incorporation of Informatics in Austrian Education: The Project "Computer-Education-Society" in the School Year 1984/85. In: Mittermeir, R.T. (ed.) ISSEP 2005. LNCS, vol. 3422, pp. 4–19. Springer, Heidelberg (2005)
4. Dagienė, V.: The Model of Teaching Informatics in Lithuanian Comprehensive Schools. Journal of Research on Computing in Education 35(2), 176–185 (2002-2003)
5. Haefner, K.: Die neue Bildungskrise. Herausforderung der Informationstechnik and Bildung und Ausbildung, Basel (1982)
6. Ershov, A.P.: Programming is a second literacy. Pocit e umela intel 1(6), 457–471 (1982)
7. Mayer, R.E.: Teaching for Transfer of Problem-solving Skills to Computer Programming. In: De Corte, E., Linn, M.C., Mandl, H., Verschaffel, L. (eds.) Computer-based Learning Environments and Problem Solving. NATO ASI Series, pp. 193–206. Springer, Heidelberg (1991)
8. Cohen, A., Haberman, B.: Computer Science: A Language of Technology. ACM SIGCSE Bulletin 39(4), 65–69 (2007)
9. Dagys, V., Dagienė, V., Grigas, G.: Teaching Algorithms and Programming by Distance: Quarter Century;s Activity in Lithuania. In: Proc. of the 2nd Int. Conference on Informatics in Secondary Schools: Evolution and Perspectives, Vilnius, November 7-11, pp. 402–412 (2006)
10. Grigas, G.: Distance teaching of informatics: motivations, means and alternatives. Journal of Research on Computing in Education 27(1), 19–28 (1994)
11. Dagienė, V.: The Road of Informatics. Vilnius, TEV (2006)
12. Clark, M.A.C., Boyle, R.D.: Computer Science in English High Schools: We Lost the S, Now the C Is Going. In: Mittermeir, R.T. (ed.) ISSEP 2006. LNCS, vol. 4226, pp. 83–93. Springer, Heidelberg (2006)
13. Bowring, J.F.: A New Paradigm for Programming Competitions. In: Proc. of the 39th SIGCSE technical symposium on Computer science education, Portland, OR, USA, March 12-15, pp. 87–92 (2008)
14. Kearse, I.B., Hardnett, C.R.: Computer Science Olympiad: Exploring Computer Science through Competition. ACM SIGCSE Bulletin 40(1), 92–96 (2008)
15. Dagienė, V.: Programming-based solution of problems in informatics curricula. In: Communications and Networking in Education: Learning in a Networked Society, IFIP WG 3.1 and 3.5, Finland, pp. 88–94 (1999)
16. Kolstad, R., Piele, D.: USA Computing Olympiad (USACO). Olympiads in Informatics 1, 105–111 (2007)
17. Dagienė, V., Skupiene, J.: Learning by competitions: Olympiads in Informatics as a tool for training high grade skills in programming. In: Boyle, T., Oriogun, P., Pakstas, A. (eds.) 2nd Int. Conf. on Information Technology: Research and Education, London, pp. 79–83 (2004)

18. Casey, P.J.: Computer programming: A medium for teaching problem solving. In: Computers in the Schools, vol. XIII, pp. 41–51. The Haworth Press, Inc., New York (1997)
19. Verhoeff, T.: The Role of Competitions in Education. In: Future World: Educating for the 21st Century: a conference and exhibition at IOI 1997 (1997) (retrieved June 2007), http://olympiads.win.tue.nl/ioi/ioi97/ffutwrld/competit.pdf
20. Pohl, W.: Computer Science Contests for Secondary School Students: Approaches to Classification. Informatics in Education 5(1), 125–132 (2006)
21. Bebras International Contest on Informatics and Computer Fluency home page, http://www.bebras.org
22. IOI home page, http://www.ioinformatics.org
23. Verhoeff, T.: The IOI is (not) a science olympiad. Informatics in Education 5(1), 147–159 (2006)
24. Verhoeff, T., Horvath, G., Diks, K., Cormack, G.: A proposal for an IOI Syllabus. Teaching Mathematics and Computer Science IV(1), 193–216 (2006)
25. Verhoeff, T.: 20 Years of IOI Competition Tasks. Olympiads in Informatics 3, 149–166 (2009)
26. Anido, R.O., Menderico, R.M.: Brazilian Olympiad in Informatics. Olympiads in Informatics 1, 5–14 (2007)
27. Kemkes, G., Cormack, G., Munro, I., Vasiga, T.: New task type at the Canadian Computing Competition. Olympiads in Informatics 1, 79–80 (2007)
28. Wang, H., Yin, B., Li, W.: Development and Exploration of Chinese National Olympiad in Informatics (CNOI). Olympiads in Informatics 1, 165–174 (2007)
29. Brodanac, P.: Regular Competitions in Croatia. Olympiads in Informatics 1, 15–23 (2007)
30. Pohl, W.: Computer Science Contests in Germany. Olympiads in Informatics 1, 141–148 (2007)
31. Casadei, G., Fadini, B., Vita, M.G.: Italian Olympiads in Informatics. Olympiads in Informatics 1, 24–30 (2007)
32. Diks, K., Kubica, M., Stencel, K.: Polish Olympiads in Informatics – 14 Years Experience. Olympiads in Informatics 1, 50–56 (2007)
33. Kiryukhin, V.M.: The modern contents of the Russian National. Olympiads in in Informatics 1, 90–104 (2007)
34. Poranen, T., et al.: Baltic Olympiads in Informatics: Challenges for Training Together. Olympiads in Informatics 3, 112–131 (2009)
35. Dagienė, V.: Information Technology Contests – Introduction to Computer Science in an Attractive Way. Informatics in Education 5(1), 37–46 (2006)
36. Dagienė, V.: Competition in Information Technology: an Informal Learning. In: EuroLogo 2005: the 10th European Logo Conference, Warsaw, Poland, August 28-31, vol. 31, pp. 228–234 (2005)
37. Dagienė, V.: The BEBRAS Contest on Informatics and Computer Literacy – Students' Drive to Science Education. In: Joint Open and Working IFIP Conf. Kuala Lumpur, pp. 214–223 (2008)
38. Opmanis, M., Dagienė, V., Truu, A.: Task Types at "Beaver" Contests Standards. In: Proc. of the 2nd Int. Conf. on Informatics in Secondary Schools: Evolution and Perspectives, Vilnius, pp. 509–519 (2006)
39. Dagienė, V., Futschek, G.: Bebras International Contest on Informatics and Computer Literacy: Criteria for Good Tasks. In: Mittermeir, R.T., Sysło, M.M. (eds.) ISSEP 2008. LNCS, vol. 5090, pp. 19–30. Springer, Heidelberg (2008)
40. Hromkovic, J.: Algorithmic Adventures: From Knowledge to Magic. Springer, Heidelberg (2009)

Impasse, Conflict, and Learning of CS Notions

David Ginat

CS Group, Science Education Department
Te-Aviv University
Tel-Aviv 69978, Israel
ginat@post.tau.ac.il

Abstract. We present a study of limited adaptation of fundamental computer science notions by computer science graduates. The examined notions involve induction, recursion and rigorous justification. We devised a problem-solving activity that revealed and addressed limited assimilation of the latter notions. The activity involved impasse phenomena, which yielded an affective reaction of conflict. The epistemic curiosity that arose from the conflict was utilized to attain insightful learning and conceptual comprehension of the above notions.

Keywords: Impasse, Conflict, Induction, Recursion, Rigor.

1 Introduction

Computer Science CS students learn a series of fundamental ideas [14] throughout their studies. These ideas are displayed and implemented in different courses, of various levels. Many of these ideas enclose a procedural facet, of actual utilization in typical occasions. For example, the fundamental idea of recursion is employed in a variety of topics, in the form of recursive algorithms. Yet, the procedural facet only embodies some of the required conception of fundamental ideas. There also is the conceptual facet [8]. This facet involves comprehension and adaptation of the ideas' diverse utilizations and inter-relations. For example, in the case of recursion, the conceptual facet involves (among other aspects) insight and utilization of the close link between recursion and induction.

We would like CS graduates to hold both procedural and conceptual understanding of the fundamental ideas they have studied. Unfortunately, this may sometimes not be the case. The understanding of ideas, or notions, may depend on how one is taught and practice the notions. Students surely see and implement the notions in straight-forward, common ways, in typical occasions. But, do they also develop suitable conceptual comprehension? Not necessarily.

In this paper, we present a study that sheds some light on this question, and relates it to learning. We examined CS graduates' conceptual comprehension and adaptation of specific fundamental notions, and noticed an undesired impasse phenomenon, of tendencies to fixate on typical utilization schemes, without fruitful outcome. This yielded a feeling of conflict, between the participants' attempted directions and their unproductive outcomes. We capitalized on the tension created due to the conflicts [1,2], and used it as a vehicle to insightful learning.

J. Hromkovič, R. Královič, and J. Vahrenhold (Eds.): ISSEP 2010, LNCS 5941, pp. 13–21, 2010.

Naive problem solvers demonstrate tendencies to fixate on an initial solution track, without diverting to alternative directions [12]. More experienced problem solvers may consider alternative directions, but still remain in a rather limited search space, which they are unable to enlarge [6]. This may occur due to limited competence in problem solving, or as a result of un-adapted epistemological obstacles [3,13]. In both cases, a feeling of momentary disequilibrium may be created [9], encapsulating a conflict between the problem-solver's cognitive structure (invoked knowledge and notions) and the environment (which requires a wider perspective) [2,4,9].

The conflict yields affective reaction [1]. Such a situation arouses epistemic curiosity [2], which encloses a strong incentive to relieve the conflict as soon as possible. This may yield an activity of critical examination of existing knowledge and tendencies, and lead one to be more receptive and rise above her current convictions [13]. Such a situation may serve as an effective ground for fruitful activity of thinking and learning [10]. When the activity is derived from some reached impasse, the problem solver tends to better remember both the situation, and its resolving notions [11].

We capitalized on the above, for revealing limited conceptions and adaptation of CS notions, and for developing suitable conceptual comprehension of these notions. In particular, we addressed the relationships between two fundamental pairs of notions that appear time and again during the CS studies: the notions of Induction & Recursion and the notions of Intuitive Argumentation & Rigorous Justification.

We conducted a study, with a group of 19 CS graduates, to whom we posed three different tasks, the first focusing on Induction & Recursion and the other two – on Intuitive Argumentation & Rigorous Justification. The natural tendencies of most of the participants were to invoke and focus on the first notion indicated in each of the above pairs. This yielded impasse phenomena, which blocked them from reaching the desired, convincing solutions. It seemed to us, that the latter mostly occurred due to epistemological-obstacle occurrences, of lacking conceptual comprehension and adaptation of the second notions in the above indicated pairs. We capitalized on the conflicts that arose due to the above impasses to teach and elaborate on the second notions, thus developing the participants' conceptual comprehension of these notions.

In the next two sections we display the impasse and conflict phenomena we experienced, with respect to the above indicated notions; and describe the consequential learning. In the last section we reflect on the study results and conclude with didactical suggestions for CS tutors.

2 Induction and Recursion

Two fundamental notions that appear throughout the CS curriculum are *induction* and *recursion*. Induction appears in proofs, algorithms (programs) design, and data-structure specifications. In proofs it offers a sound justification means; in algorithm design it appears in incremental modifications; and in data-structures it is used in structure specifications. Recursion appears similarly in proofs, algorithm design, and data-structures; but in "reverse" specifications.

Thus, both induction and recursion encapsulate an incremental motive, but in induction the increment is viewed and specified forward, whereas in recursion it is viewed and specified backwards.

The close link between the two notions yields the relevance, and possibly utilization, of one of them when the other is introduced. For example, the notion of dynamic programming often involves recursive rules that are implemented with an inductive, forward computation [5]. Alternatively, a task or data-structure that is originally specified inductively may be implemented recursively.

Suitable conceptual understanding of these two notions involves comprehension and recognition of the above relationships, including the transformation of the train of thought from one of them to the other. We would like CS graduates to be aware of these possible transitions, which are beyond the primary, straightforward utilizations of each of these notions.

Thus, we posed the following task to our study participants in order to examine and address their comprehension and recognition of the above.

Binary Sequence. A sequence of words is defined in the following way: $W(1)=0$; $W(2)=001$; and each successive word is obtained by substituting a 0 in the previous word by 001 and a 1 – by 0 (thus, $W(3)=0010010$). Develop a computer program for which the input is a positive integer N, $1<N<1,000,000,000$, and the output is the value of the Nth bit in the first word whose length is equal-to or greater-than N.

The above task involves an *inductive* definition of a sequence of words. The initial, straightforward direction of approaching the task is by generating the words, one-by-one, according to the given inductive rule, until a long-enough word is obtained. This solution requires an array, which will be repeatedly updated and keep the currently generated word. A majority of the participants initially followed this direction.

Yet, N may be very large, and it is impossible to use an array of size 1,000,000,000. Most participants realized the difficulty, and sought a different approach. Some tried to recognize some recurring pattern in the linearly generated sequence, while others tried to "save" some memory in various other ways. Their outcomes were either erroneous (e.g., faulty general patterns) or still unacceptably inefficient, space-wise. They were all fixated on the forward point of view expressed in the task's specification, with the inductive rule of word generation. They seemed to be "blocked" by the impasse of literally simulating the task's forward rule. At this point, we capitalized on the (frustrating) conflict that occurred between their unsuccessful attempts and the required results.

The main obstacle in this task is to somehow avoid generating all the sequence words, one after the other. Perhaps a switch in the direction of the train of thought, from forward to backwards, may help. So, it is worth examining whether the inductive sequence definition in the task's specification may be reformulated into a recursive rule. Apparently, such reformulation is possible. After examining some small examples, one may notice that it is possible to specify a recursive rule that encapsulates the decomposition of a sequence word into three concatenated parts, as follows:

$$W(k)=W(k-1)W(k-1)W(k-2)$$

That is, the current word, $W(k)$, is the concatenation of $W(k-1)$ with $W(k-1)$ and then $W(k-2)$. One may easily see this with $W(3)$, and then generalize, and argue this observation inductively.

Once one realizes the above concatenation, one may avoid the actual generation of all the sequence words. In order to compute the value of the Nth bit in the first word whose length is equal-to or greater-than N, it is sufficient to record the *length* of each word. This can be done using a look-up table, with a single entry for each word. Since the length of the words grows exponentially, their number is very small, and so is the size of the required table.

Each entry in the look-up table may be constructed inductively based on the values of the previous two entries (in a dynamic programming manner). Then, the value of the Nth bit may be computed recursively. The total computation time and space is of O(logN).

We illustrate the above with a short example. Let N=20. The algorithm will first *inductively* construct the look-up table, L, of the lengths of the sequence words: L(1)=1 (the length of W(1)); L(2)=3; L(3)=7; L(4)=17 (twice L(3) plus once L(2)); L(5)=41.

Then, the algorithm will conduct the *recursive* computation. W(5) is the first word of length equal-to or greater-than 20. Upon applying the recursive rule on W(5), the algorithm will find that the 20th bit in W(5) is actually the 3rd bit in W(5)'s second W(4) component. (Recall that W(5)=W(4)W(4)W(3), and L(4)=17). So now the problem is reduced to the computation of the value of the 3rd bit in W(4); and so on. Eventually, the computation will yield the value 1 (for the 20th bit in W(5)).

Upon reflecting on the whole solution process, the study participants realized their impasse, and comprehended the important role of switching the train of thought, from forward to backwards. They also noticed that the final solution encapsulated both inductive and recursive applications (of the devised recursive rule) – induction for constructing the look-up table, and recursion for computing the desired bit value (based on the table). They indicated that the above example illuminated to them a facet of the relationships between induction and recursion of which they were not fully aware.

3   Intuition vs. Rigor

The previous section displayed an impasse phenomenon that disabled the algorithm designers from reaching a suitable algorithm. Sometimes, algorithm designers attempt an algorithm-design task for which they reach a suitable solution and conjecture its correctness; but they are unable to provide a solid, convincing correctness argument. In addition, sometimes CS problem solvers are required to analyze and assert the primary characteristics of a given algorithm. They provide an intuitive conjecture; yet, they, again, are unable to offer a rigorous justification. We examined and addressed these two cases with two different tasks. We first display our experience with the former case, and then present our experience with the latter one. We posed the following task, for illuminating convincing termination justification.

Line Switching. A matrix of N×N integers, of positive and negative values, is given together with the operator Switch, which may be applied on a single row or a single column at a time. An application of Switch on a line (row/column) switches the sign of each of the integers in that line. The goal is to reach a matrix in which the sum of the values in each line (row/column) is not negative. Develop an algorithm that inputs a given matrix and obtains the goal if it is reachable (and notifies that the goal is unreachable otherwise).

Example. The figure below illustrates a single application of Switch, on the top row.

-3	7	9	-6	-8
-4	-6	-7	-8	9
9	-9	7	-5	7
-5	9	-8	3	6
-8	5	0	9	-7

→

3	-7	-9	6	8
-4	-6	-7	-8	9
9	-9	7	-5	7
-5	9	-8	3	6
-8	5	0	9	-7

The solution seemed rather straightforward to many of the participants: repeatedly seek a line with a negative sum, and operate Switch on that line. Some participants offered to always seek the line with the minimal sum. Indeed, this algorithm seemed to always terminate successfully. But, how can this be argued in a convincing why? The participants seemed to face an apparently familiar, but unfortunate difficulty.

The operation of Switch on a particular line "takes care of" this line; but may "distort" some other line. This may be seen in the example above: the switching of signs of the first row changes the sum of that row from negative to positive; but it also "distorts" the middle column. Thus, progress is not that obvious. Students are used to show progress by locally looking at the data on which the last operation was performed. This is the case in many, simple and complex algorithms.

Yet, it is not the case here. Simply looking at the line on which Switch is operated does not reveal progression. Some participants tried to argue that the value of the line with the minimal sum in the matrix always increases; but this is not always true. They were fixated on a <u>local view</u> of the operator's application, which did not yield a convincing argument. This raised a conflict between the intuitive feeling of correct design and the inability to rigorously justify correctness.

We turned to resolve the conflict, by seeking together with them a suitable *metric* with which to argue progress. Various attempts were made by trying diverse subsets of the matrix, as elements on which to focus. Eventually they realized that it is beneficial to look at the sum of the values of all the matrix integers. A single application of Switch increases that sum (as it transforms the sum of the line on which it is operated from negative to positive). Since the total sum that may be reached is bound (by the sum of the absolute values of the matrix integers), the algorithm progresses, and will eventually reach the desired goal for any initial matrix.

In reflecting on the solution process, the participants realized the importance of seeking a metric, with which rigor may be argued convincingly. They also became more aware of the notion of "eventuality", in the sense that perhaps one cannot tell the exact number of operations that will yield the goal, but one can clearly tell that the goal will be reached.

In examining and analyzing the characteristics of a given algorithm, the natural tendency is to view the algorithm's execution through an operational perspective, of its various execution "behaviours". This may, unfortunately yield only "seemingly convincing" arguments. Rigorous arguments may likely be obtained from a different, assertional perspective, which embeds assertions of characterizing patterns [7]. Unfortunately, many follow the former, unsatisfying operational perspective, rather than the latter assertional one. We examined and addressed this with the following task.

Board Staining. The operator **Stain** is repeatedly applied on a board of N×N cells, of which some are initially stained. A single application of **Stain** on the board results in the staining of all the clear cells adjacent to two or more already-stained cells. Is there an initial configuration of N-1 stained cells for which repeated application of **Stain** will result in a completely stained board? Justify your answer.

Example. The figure below illustrates a single application of **Stain**.

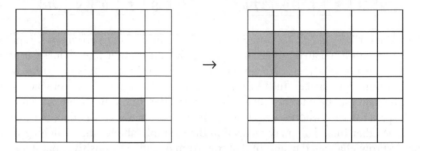

In the above example, further applications of **Stain** will yield the staining of most, but not all of the board cells (the top line, bottom line, and right column will remain clear).

The study participants repeatedly tried the **Stain** operator and learnt its behaviour through various invocation sequences. After examining diverse cases of initially N-1 stained cells they conjectured that whole-board staining may be impossible.

Further examination of the operator led quite a few participants to the observation that K stained cells, which are diagonally placed next to one another, lead to the staining of their corresponding square of K^2 cells (in K-1 **Stain** invocations).

The figure below displays staining starting from a "diagonal start".

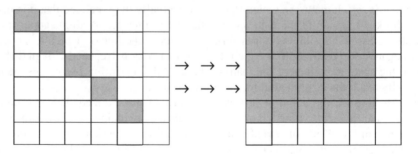

The "diagonal-start" above looked to the participants like a configuration that yields <u>maximal</u> number of stained cells, in the sense that no other initial configuration with the same initial number of stained cells may yield larger staining. They argued that no configuration of N-1 stained cells will lead to a whole-board staining, since the maximal staining structure, of "diagonal start" only yields $(N-1)^2$ stained cells.

The latter argument makes sense and seems intuitively right. But, is it convincing? While one may indeed see the rational behind it, one may not be fully convinced. The "maximum" conjecture of "diagonal-start" should be rigorously justified. That is, one should show that <u>any</u> other initial structure does not stain an area larger than the one stained by "diagonal start". The study participants tried to characterize the behaviours of staining sequences, but were unable to obtain a convincing argument. They were fixated on examining the operational behaviour of diverse starting points (and comparing them to "diagonal start"), without carefully seeking and unfolding a rigorous characteristic on which to capitalize. This, again, yielded a feeling of frustrating conflict, as they were only able to offer intuitive arguments, without rigorous justification.

They were directed to pursue a different approach, of carefully observing the operator's characteristics through a <u>single</u> staining step, rather than a sequence of steps. In this examination, they were recommended to seek a particular type of assertional pattern – an invariant property.

The sequence of staining steps adds stained cells. Thus, at first glance it seems natural to examine invariance with respect to the number of added stained cells. A single invocation of **Stain** does not add more than twice the number of already-stained cells. Can one capitalize on this observation? It seems that this straightforward observation will be of little help. One should seek more hidden patterns.

Perhaps, a feature other than the number of added stained cells can be relevant. The change in the stained area is not only in its size but also in its shape. Global shape alternation seems too complex to analyse, but local alternation may be rather simple to examine. The two <u>basic shapes</u> that yield further staining are displayed in the figure bellow.

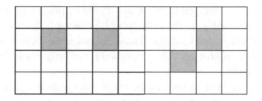

A basic shape feature that comes to mind is circumference. For each of the two basic cases in the above figure, the circumference of the stained area does not increase by the application of **Stain**. The explanation for this is that in order for a new cell to become stained, at least two of its sides must be a-priori on the circumference of the stained area; and when it is stained, these sides do not remain anymore on the circumference of the stained area. Thus, the contribution of this newly stained cell is that at least two sides are removed from the stain circumference, and at most two sides are added. This implies a powerful, global invariant:

Staining Invariant. *The circumference of the stained area does not increase by an application of* **Stain**.

The circumference of any N-1 cells is at most 4×(N-1). The circumference of the whole board is 4×N. Thus, whole-board staining may never occur, as we know from the invariant that the stain circumference never grows.

Looking back at the solution process the participants noticed that their initial "sensing" of the problem was obtained through the operational perspective. Observations of series of invocations yielded the conjecture of "maximal" structure. Yet the participants realized that although their operational perspective paved the initial way, it fell short of providing a convincing, rigorous justification.

Rigor was only obtained after invoking and following the assertional approach, and seeking an illuminating invariant. The revealed invariant was then tied to the initial state and the required final state, in the construction of the final rigorous argumentation.

4 Conclusion

The last two sections presented our study of impasse, conflict, and learning, with the pair of the notions of Induction & Recursion and the pair of the notions of Intuitive Argumentation & Rigorous Justifications. The study participants seemed to be fixated in the first notions of these pairs, upon attempting to solve the three tasks that were posed to them. These fixations yielded frustration and conflict, due to the unsuccessful outcomes reached by the participants' solution attempts. We capitalized on this affective phenomenon to elaborate on the significant relationships between Induction and Recursion, and the essential role of the notions of invariance and metric in yielding rigor.

In a discussion that followed the problem-solving activity, the participants, who completed three or four years of CS studies, indicated to us that they recall "seeing some relationship between induction and recursion, and some utilization of invariants and metrics". Yet, they did not seem to conceptually comprehend and adapt these notions and their inter-relationships. In particular, they adapted little of the induction-recursion link, and expressed a very limited level of justification, primarily with intuitive conjectures. The participants acted as if the conceptual facets of induction-recursion and rigor were epistemological obstacles to them. They did not seek these notions or considered them during their solution process. They pursued them only under our careful guidance and illuminating direction.

The reflection that we held in the end of each task's solution process illuminated two elements to the study participants: the limited conceptions they held prior-to and during the solution process, and the essential role of the notions and relationships which paved the way to the suitable solutions. Some of the participants indicated that they felt that our intervention and guidance "opened a new perspective" and developed their conceptions, in a manner that will be remembered, as it occurred in situations in which they felt "stuck" and unable to advance with the notions they have accumulated during their CS studies.

We believe that the approach presented here may contribute an additional means for developing students' conceptual understanding and adaptation of fundamental CS notions. Students usually learn and assimilate procedural facets of studied notions before they realize and comprehend conceptual facets. The latter ones develop only when one sees diverse utilizations of the studied notions and their inter-relations. This requires examples which may sometimes be rather different from the typical (and sometimes routine) ones that students see and practice in their various courses.

Tutors may identify notions with which their students demonstrate limited conceptual understanding. They may then devise activities, such as the one presented here, in which the students may be fixated on unsuitable, ineffective solution directions. The affective reactions to the experienced impasses may lead the students to critical examination of their own directions, raise curiosity, yield fruitful learning and elaborated awareness of the conceptual facets of the addressed notions.

References

1. Anderson, L.W., Bourke, S.F.: Assessing Affective Characteristics in the Schools. Lawrence Erlbaum Associates, Mahwah (2000)
2. Berlyne, D.E.: Conflict, Arousal, and Curiosity. McGraw-Hill, New York (1960)
3. Brousseau, N.: Theory of Didactical Situations in Mathematics. Kluwer Academic Publishers, Dordrecht (1997)
4. Brown, S.A.: Exploring epistemological obstacles to the development of mathematics induction. In: Proc of the 11th for Research on Undergraduates Mathematics Education (2008)
5. Cormen, T.H., Leiserson, C.E., Rivest, R.L.: Introduction to Algorithms. MIT Press, Cambridge (1991)
6. Dempster, F.N., Corkill, A.J.: Interference and inhibition in cognition and behaviour: unifying themes for educational psychology. Educational Psychology Review 11(1), 1–88 (1999)
7. Dijkstra, E.W., et al.: A Debate on Teaching Computing Science. Communications of the ACM 32(12), 1397–1414 (1989)
8. Hiebert, J., Lefevre, P.: Conceptual and procedural knowledge in mathematics: an introductory analysis. In: Hiebert, J. (ed.) Conceptual and Procedural Knowledge: The case of mathematics, pp. 1–27. Lawrence Erlbaum Associates, Mahwah (1986)
9. Mischel, T.: Piaget: cognitive conflict and the motivation of thought. In: Mischel, T. (ed.) Cognitive Development and Epistemology. Academic Press, London (1971)
10. Movshovitz-Hadar, N., Hadas, R.: Preservice education of math teachers using paradoxes. Educational Studies in Mathematics 21, 265–287 (1990)
11. Patalano, A.L., Seifert, C.M.: Memory for impasses during problem solving. Memory & Cognition 22(2), 234–242 (1994)
12. Schoenfeld, A.: Learning to Think Mathematically: Problem Solving, Metacognition, and Sense Making in Mathematics. In: Grouws, D.A. (ed.) Handbook of Research on Mathematics Teaching and Learning, pp. 334–370. Macmillan, Basingstoke (1992)
13. Sierpinska, A.: Humanities students and epistemological obstacles related to limits. Educational Studies 18(4), 371–397 (1987)
14. Schwill, A.: Fundamental ideas of computer science. Bulletin of European Association for Theoretical Computer Science 53, 274–295 (1994)

K-12 Computer Science: Aspirations, Realities, and Challenges

Allen B. Tucker

Bowdoin College

Abstract. Now more than ever, revitalization of computer science as a mainstream discipline within K-12 education is an important goal worldwide. Especially in the US, where most states fail to understand the nature of computer science as an academic field, the challenges are enormous. What recent efforts have been made in the US toward reaching this goal, and to what extent have these efforts been successful? This paper addresses these questions.

1 Aspirations

In 2003, many were concerned that computer science was not well-understood by the general public in the United States, nor were its concepts and skills viewed as particularly important.

To help address this concern, the ACM sponsored the development of a "Model Curriculum for K-12 Computer Science" (1). The Model Curriculum provided a framework within which:

- individual states could develop academic standards for computer science,
- school districts could implement new courses that taught computer science principles to a wide range of students,
- new teaching materials could be developed to support these new courses, and
- schools of education and in-service programs could prepare teachers to offer these new courses throughout K-12.

If these goals were met, a majority of high school graduates would become better informed about the concepts, techniques, and applications of computer science in the modern world, and thus would become more effective citizens in the 21st century.

1.1 The ACM K-12 Model Curriculum

The Model Curriculum was expressed in four levels, designed for use in grades K-8, 9-10, 10-11, and 11-12 respectively, as summarized in Figure 1. Below is a summary of the curriculum content at each of the four levels.

J. Hromkovič, R. Královič, and J. Vahrenhold (Eds.): ISSEP 2010, LNCS 5941, pp. 22–34, 2010.

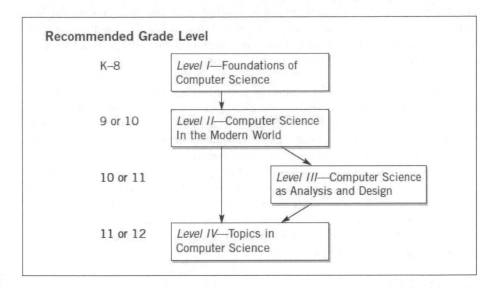

Fig. 1. The ACM K-12 Model Curriculum

Level I (grades K-8): Foundations of Computer Science. This curriculum provides elementary school students with foundational concepts in computer science. It integrates basic skills in technology with simple ideas about algorithmic thinking. Hands-on activities help ensure that students meet these goals. For examples of such activities, see **http://csunplugged.org**.

Level II (grade 9 or 10): Computer Science in the Modern World. This course is designed to be accessible for all students, whether they are college-bound or workplace bound. It includes fundamental concepts of computer operations (hardware, software, operating systems, etc.), computer networks, the Internet, and algorithmic problem-solving. It also exposes students to computing careers and ethical issues.

Level III (grade 10 or 11): Computer Science as Analysis and Design. This is a one-year elective course that should earn a math or science credit. It continues the study begun at Level II, but emphasizes the scientific and engineering aspects of computer science, focusing on mathematical principles, algorithmic problem-solving and programming, hardware, networks, and social impact. The Computer Science AP course can be part of this level.

Level IV: Topics in Computer Science. This can be either a projects-based course or a course leading to industry certification. In either case, the Level II course is a prerequisite. Example projects-based courses include Multimedia, Graphics, Web Site Development, Animation, Networking, Simulation and Modeling. Example industry certification courses include Certified Internet Webmaster (CIW), A+ Certified Technician, and i-Net+.

The K-12 Curriculum also described a number of challenges that would need to be faced in order for computer science to become widely taught in the US. At that time, the only coherent computer science activity at the K-12 level was the AP curriculum, which reached only a small minority of gifted high school students.

At the time the K-12 Curriculum was published, state-level academic standards for K-12 computer science were nearly nonexistent, as was teacher training in computer science (outside of the AP curriculum). However, Information Technology (IT) was part of the curriculum, but was taught as a skill needed to support the traditional academic subjects, mainly science.

So beginning in 2004, a lot of work needed to be done if computer science were to be established as a regular discipline at the K-12 level throughout the US. With 50 different states each setting its own academic standards, this was (and still remains) a particularly daunting challenge. However, it is also clear that widespread state acceptance if computer science as a core discipline is necessary for preparing young people for life in the 21st century.

1.2   The CSTA

To help with this work, the ACM sponsored the development of a new Computer Science Teachers Association (CSTA). CSTA is a semi-autonomous organization with its own Steering Committee, membership and distinct member benefits under the auspices of ACM. From its founding in 2004 to the present, CSTA membership has grown to over 7,000 high school and middle school teachers, college and university faculty, and industry representatives.

The purpose of CSTA is to support and promote the teaching of computer science by helping to build a strong community of CS educators, providing opportunities for high quality professional development, advocating for a comprehensive K-12 computer science curriculum, supporting projects that communicate the excitement and opportunities of CS, conducting research about computer science education, and providing policy recommendations to improve the status of CS in the high school curriculum

2   Realities

Recent reports by the Bureau of Labor Statistics shows Computer Science as the fastest-growing professional sector for the decade 2006-2016. Percentage increases in the numbers of network systems analysts, computer software engineers, and computer systems analysts are predicted to be 53%, 45%, and 29% over that decade, while increases in the numbers of Biological scientists, chemists, electrical engineers, and mechanical engineers are all predicted to be less than 10% during the same period (2).

At the same time, overall US progress between 2004 and 2009 toward reaching the aspirations expressed by the Model Curriculum has been glacial.

- individual states have been slow to understand or develop academic standards for computer science,

- only a small number of school districts have implemented new courses that teach computer science principles for a wide range of students,
- only a modest amount of new teaching materials has been developed to support these new courses, and
- schools of education and in-service programs have not trained significant numbers of teachers to offer such courses throughout K-12.

The major progress being made for K-12 computer science seems to be taking place at the grass-roots level, which we shall discuss in more detail in Section 3.

2.1 Student Enrollment Levels

From 2005 to the present, the CSTA has conducted surveys (5) of all 14,000 school districts in the US to determine the degree to which computer science in any form is being taught. Here is a summary of the results of its 2009 surveys:

Responses were received from 1153 of all the school districts surveyed. This 8% response level is a strong indicator of the widespread indifference or ignorance about computer science that persists throughout the public school system. Of these responses, 65% reported that their school offered one or more introductory (pre-AP) computer science courses. However, only 44% reported that the course was required for all students.

Looking further at these responses, we learn that the content of what is called "computer science" at the K-12 level is dominated by information technology skill-building — teaching students how to use computers in support of the traditional disciplines. For example, 56% of the schools teach computer science as a business or "tech" credit, rather than a math or science credit.

2.2 State-Level Academic Standards

In spite of the K-12 Model Curriculum and subsequent efforts by the CSTA, a widespread lack of urgency seems to persist among the 50 states about the importance of computer science in the K-12 curriculum.

Most individual state academic standards indicate that computer science is identified as IT and it typically shows up under the heading "Science and Technology." However, there is great variation among different state standards about the place of computer science in the K-12 curriculum. Here are a few examples that indicate the flavor of current state standards:

Virginia. Computer Technology Standards of Learning for Virginia's Public Schools, Board of Education – Commonwealth of Virginia, (June 22, 2005): "The Computer/Technology Standards of Learning identify and define the progressive development of essential knowledge and skills necessary for students to access, evaluate, use and create information using technology."

Pennsylvania. Academic Standards for Science and Technology, Pennsylvania Department of Education (January 5, 2002) focuses exclusively on the use of technology in science education and nowhere else.

New Jersey. (2009) "In grades 9-12, students demonstrate advanced computer operation and application skills by publishing products related to real-world situations (e.g., digital portfolios, games and simulations), and they understand the impact of unethical use of digital tools. They collaborate adeptly in virtual environments and incorporate global perspectives into problem solving at home, at school, and in structured learning experiences."

North Carolina. Computer/Technology Skills Standard Course of Study and Grade Level Competencies K-12 Department of Public Instruction (Revised 2004): "The Computer/Technology Skills Standard Course of Study describes the progressive development of knowledge and skills in six strands: Societal and Ethical Issues, Database, Spreadsheet Keyboard Utilization/Word Processing/Desktop Publishing, Multimedia/Presentation, and Telecommunications/Internet"

California. Career Technical Education (May 2005): Information Technology Industry Sector identifies 4 Career Pathways: Information Support and Services, Media Support and Services, Network Communications, and Programming and Systems Development.

Florida. Teacher Certification Examinations (FTCE) are standardized tests used to assess the competencies of prospective teachers according to Florida's Sunshine State Standards. FTCE refers to 47 different exams: four General Knowledge sub-tests, one Professional Education exam, and 42 Subject Area examinations. Computer science is one of the 42 subject area examinations.

Thus, the overriding emphasis within state standards is upon teaching students to use technology to support learning in other academic disciplines, rather than presenting computer science as a distinct discipline of its own.

2.3　Public Confusion

Ignorance about the discipline of computer science among state departments of education reflects a wider ignorance by the general public about the nature, methodologies, and contributions of computer science in the modern world. The following types of misperceptions about computer science tend to dominate the public discourse:

CS = programming
CS = computer literacy
CS = a tool for studying science
CS = IT
CS is just for white males
CS is not a scientific discipline

Unfortunately, these misperceptions often drive K-12 policy decisions about computer science funding in the face of budget shortfalls and competing priorities. Here is an interpretation of recent discussions by the Kansas Board of Education, which illustrate the level of public confusion:

August 26, 2009

Recently the Board convened a task force to review the Qualified Admissions Regulations, which concluded that the technology requirement is outdated and that the content is being taught in other courses. Based on this conclusion, the Board is proposing to cut the computing technology requirement.

It turns out that while the technology requirement was intended to be a basic computing literacy course, it allowed many high schools to develop courses with computer science content. ACM and CSTA's concern is that if the Board eliminates the computing technology requirement students will focus only on the core requirements and computer science in Kansas will disappear.

Cameron Wilson
ACM Director of Public Policy

2.4 Equity Issues

A more subtle outcome of the widespread public confusion about computer science is that course offerings systematically prevent access by a significant pool of young talent – women and members of ethnic minorities. Here is a recent statement by Barbara Ericson, Director of CS Outreach at Georgia Tech, that characterizes this situation:

> Georgia Tech just received a NSF grant to retrain unemployed IT workers to be high school computing teachers and pair them up with existing computing teachers during the first year of teaching. So, my co-teaching this course helps me see how this will work.
>
> There are 28 students in the class. As we often see in an AP CS course the majority are white and male. There are nine females and three African Americans in the course. This course is in marked contrast to the Business Essentials course just before it that is 75% African American. (The school is actually very diverse with 41.2% White, 35.3% African American, 12.3% Hispanic, 11.5% Asian and other racial groups.)
>
> One of the things I would like to do this year is recruit a more diverse class for next year. I also would like the school to offer at least two sections for AP CS next year. I expect demand for AP CS to increase greatly in Georgia in the next few years since it counts as one of the four years of science starting with students who were freshman in 2008-2009.

The good news here is that the State of Georgia is counting AP Computer Science as one of the four years of science required of all students. The bad news is that there is no institutional guarantee that this kind of course will be accessible to significant numbers of young women or ethnic minorities.

2.5   Elective vs. Core Subject Matter

The term STEM is a widely-used NSF term that characterizes the combined fields of "Science, Technology, Engineering, and Mathematics." How and whether computer science fits into STEM isn't always clear, particularly in the K-12 education discussion.

For example, computer science can be viewed as either a technology, math, or even business credit depending on the institutional setting. Moreover, scientists often confuse and intermix the terms "technology literacy" and "computer science" in their discussions, as if to say the two are the same. Thus, computer science is easily relegated to the role of a supporting skill in service to the mainstream mathematics, science, and engineering fields.

At the K-12 level, this view translates to the notion that computer science is an elective rather than a core subject. When offered as an elective, computer science is becoming increasingly difficult for students to fit into their already-crowded schedules. This situation is made worse by the current trend to increase the number of math and science courses required for graduation. (With few exceptions, computer science is usually counted as neither.)

The result of this policy is a further decline in student awareness and interest in computer science. For example, a 2008 New Jersey survey found that the percentage of college-bound students who said they were interested in majoring in computer and information science dropped steadily from 4.5% to 2.9% during the period 2003-2008.

2.6   CS Enrollments and "No Child Left Behind"

The policy called "No Child Left Behind" (NCLB) seems to have inadvertently left computer science behind, forcing school districts to take resources away from electives and put them into remedial education. Adopted by Congress in 2002, NCLB requires every state and school district to:

- apply standardized tests for all students in Math, Reading and Science
- measure schools adequate yearly progress in Math and Reading
- assign highly qualified teachers to teach the core courses
- form Math and Science partnerships focused improving teaching through development, curriculum and pedagogy research
- provide grants to states and local districts to use technology in the classroom
- require students be technology literate by 8th grade

Since its inception, NCLB has had three practical impacts on Computer Science. First, its focus on core course and testing requirements means that electives like computer science become starved for resources. Second, its focus on technology literacy further diminishes and confuses the importance of computer science education and the critical thinking skills that it uniquely embodies. Third, because of this confusion, there is no mandate to train teachers to teach computer science rather than technological literacy.

2.7   The AP Conundrum

Due to a variety of factors, enrollments in the Advanced Placement (AP) computer science exams have dropped by 15% during the period 2002-2008, while enrollments in AP exams in the other sciences and mathematics have grown significantly during the same period. Particularly discouraging is the low participation rate by females (17%) and ethnic minorities (11%) in AP computer science.

In 2008, the College Board announced the cancellation of the Advanced AP Computer Science Exam (The regular exam is still administered). The Board has pledged to place more energy on the development of the regular AP exam, but the content of that exam is limited by the narrowness of the introductory computer science course offered at most universities.

The way we teach computer science in colleges, and hence the narrow emphasis of the AP exam on programming, contributes to the many widespread public misconceptions about the nature and importance of computer science in the world, as documented above.

2.8   Teacher Training and Certification

Due to the absence of state curriculum standards in computer science, teacher preparation in computer science among schools of education is nearly non-existent. Many states have no requirements at all, so anyone can teach computer science. Moreover, most teachers and administrators don't understand the computer science requirements in their own states, to the extent that they exist at all. Computer science AP courses are usually staffed by math, science, and vocational education teachers.

3   Challenges and Opportunities

Despite the realities of the last five years, many individuals and groups have been working at the grass-roots level to help change the landscape for K-12 computer science. While most of these efforts are taking place at the grass roots level (often with support from the CSTA), they are impressive and they may ultimately turn the tide of K-12 computer science in a more constructive direction.

3.1   Grass-Roots Efforts

To characterize the intensity with which many high school teachers are working at developing computer science in their schools, I recently received the following letter from a high school teacher in Bangor, Maine:

> I am currently working on improving the focus my high school has on technology in the curriculum. Currently we require only a single quarter credit (9 weeks) of instruction for graduation, and this during the freshman year. As you might guess this is woefully inadequate.

I currently teach three different courses at the high school level in computer science, ... two introductory computer science courses ... and a selected topics course, where students are working on a self determined project under my guidance. Previous projects have involved creating and programming device interfaces, creating a client server program, writing educational software used in introductory programming classes.

I am trying to move our program from a purely elective program to one which requires credit for graduation in the area or technology innovation (computer science would be one example). ... I see education at the 9-12 level as being central in promoting more study at the college level in fields such as computer science, or at the very least fostering an appreciation for the technology that has become so central to everyone's lives.

I am looking for information which will help to move our school's faculty in that direction. I am a member of the CSTA, and I have found a number of good resources there as well. Do you know of any collaboartives in the state on Maine that are working on this issue? Do you know of any other schools in the state or even the country at the 9-12 level that are working on a similar set of goals?

My reply to this letter encouraged the writer to continue networking, especially through the CSTA, to find colleagues who are working toward the same end. In a larger sense, this letter typefies the frustrations of many teachers who are trying to develop computer science in their schools; they often work in isolation and against enormous barriers of institutional indifference and ignorance.

3.2 CSTA Efforts and Impact

This year, CSTA reports (6) that more than 25% of secondary schools have implemented the ACM Model Curriculum in computer science. It also reports progress in all of the following activities:

- CSTA is working with the College Board and the National Science Foundation to create a new AP CS course that will be both rigorous and engaging for all students Curriculum
- 90 CSTA workshops have been conducted to help teachers improve both their knowledge and their teaching skills in computer science
- CSTA careers brochures and posters are in every secondary school in the country.
- The CSTA Leadership Cohort project is providing leadership and outreach training to two teacher leaders from each state, creating a cohort of master teachers who are working for improvements at the local level.

3.3 Teacher Development Institutes

Several initiatives have been taken by universities and collaboratives to help provide in-service training for K-12 computer science teachers.

For example, the Institute for Computing Education at Georgia Tech is working to improve the quality and quantity of AP CS teachers, as well as increase the number and diversity of computer science students in Georgia.

Another such initiative is called the "First Bytes Collaborative Workshop," offered by the University of Texas at Austin. Its fourth annual offering will take place in July 2010. The goals of the workshop are to:

- Collaborate to improve Computer Science education in Texas.
- Learn about exciting new technologies that are impacting student learning.
- Share information on recent trends in Computer Science education at the University level.
- Provide recent data on employment opportunities for students with Computer Science degrees.
- Work in a team.
- Exchange experiences and effective teaching methods among colleagues.
- Build relationships between Computer Science high school teachers and UT-Austin faculty.

3.4   On-Line Communities and Teaching Materials

Other collaborations are taking place in an on-line setting. For example, the Computer Science Unplugged program (csunplugged.org) supports a Google discussion group (http://groups.google.com/group/cs-unplugged-sharing) to encourage the sharing of computer science teaching materials and experiences among K-12 teachers.

The CMU CS4HS web site http://www.cs.cmu.edu/cs4hs provides another example of on-line K-12 teacher communities forming to share computer science teaching materials and experiences.

The CSTA Source web repository is a searchable database with hundreds of unique resources: lesson plans, modules, and presentations for K-12 teachers. Every resource is reviewed by a committee of experts to ensure that it is complete, relevant, appropriate, and pedagogically sound. The classification system links directly to the ACM K-12 Model Curriculum, so that teachers know what resources are available for each course in the curriculum.

The importance of on-line communities to the grass-roots development of computer science curriculum and pedagogy is recognized in a recent letter from a member of the CSTA Board of Directors:

> The more I see and think of these issues, the more I realize how important it is that we are clear in our message, and how the Level I, II, and III courses in the ACM Model Curriculum present that message. Computing technology is everywhere. ... But it's hard to convey this breadth to prospective college students. I keep going back to the nature of the Level II course, which would cover some of the major applications to which computing is put, and then look a little deeper into the technology necessary to make those applications actually work.

And this takes me back to programming the Android phone, a phone with a complete browser capability built in, and that can also be programmed for games. It's very slick, and to make this work there must be a great deal under the hood that the better students, who might become computer science majors, need to be aware of even if they never actually program an Android. It is enough that they know of and understand the existence of these layers of software that distinguish a modern mobile device from a paperweight.

3.5   Humanitarian FOSS

To help enable better student understanding of such technologies as the Android phone, some vehicle for providing student access to these developments.

Recently, a new area of computer science has emerged called Humanitarian Free and Open Source Software (FOSS for short) development. This is a new way of teaching software development that includes student engagement in projects that develop FOSS artifacts for humanity. The development of software for the Android phone mentioned above is one such project.

It is currently estimated that open source software usage has increased its footprint to more than 20% of all desktop computers worldwide (4). However, the nature of open source software, together with its distinctions from proprietary software, is not well-understood by the general public, let alone many persons in the computer science field itself.

There are two ways in which K-12 students can engage FOSS as a new part of computer science. One way would add the study of open source software and the unique nature of its licensing in a teaching unit on intellectual property and computer ethics. The other would engage students directly in the process of developing open source software, such as the software underlying the Android phone. With this, students would learn about team programming, participation in discussion threads, reporting and fixing bugs, and developing user documentation.

K-12 students are especially well-prepared to participate in the FOSS development process, since they are already quite savvy about participating in on-line social networks (Facebook, Twitter, etc.). Adding FOSS concepts and practices as a new element in levels II, III, and/or IV of the ACM Model Curriculum would not only help modernize the curriculum but would also make computer science more attractive to the many bright students who are otherwise inclined to favor other disciplines.

The Humanitarian FOSS movement has many ongoing projects that are quite accessible to high school students. Summer institutes sponsored by colleges, corporations, and other groups testify to the dynamic growth that is occurring in this area. At Trinity College, for instance, the H-FOSS summer institute has students working on such projects as the Sahana Disaster Relief software, the OpenMRS medical records system, and the Androd phone project. For more information about this institute, see `http://hfoss.org`.

3.6   Pair Programming

An essential part of open source software development is the notion that students contributing to large software projects must work in teams of two or more rather than in isolation. Talking about the benefits of pair programming in classroom exercises, one instructor puts it this way:

> Overall, the benefits of higher confidence levels and more concerted efforts to develop a solution before turning for help vastly outweigh the occasional inequities that occur within groups. On a selfish note, it also means half the number of labs to grade! If you haven't tried pair programming in your class, you might try it on just one lab or assignment to see how it goes.

3.7   Post-secondary Computer Science

In many ways, the elephant in the closet for K-12 computer science is what is going on in many colleges and universities, especially in the first course. The fact that the first course in computer science is only a programming course sends a lot of wrong messages to students who have no idea about the nature of computer science itself. As a result, this course may be turning away a lot of bright students who migrate to mathematics, economics, or science majors because it offers a more interesting and challenging field of study.

To corroborate this view, computer science suffers an unusually low retention rate for students who continue into the major. Some studies show this rate at 38%, which is far worse than engineering, business, social science, and other science majors. These studies span over 17 years, so the low retention rate cannot be attributed just to the dot com bust.

So the suggestion here is that computer science curriculum reform continues to be needed at the collage and university level, as well as at the K-12 level. For a discipline that is changing so rapidly, curriculum reform remains an especially difficult challenge at all levels.

4   Conclusions

In 2003, many were concerned that computer science was not well-understood by the general public in the United States, nor were its concepts and skills viewed as particularly important. In 2004, the ACM K-12 Model Curriculum was designed to help change this situation. Events in the last five years indicate that such change is gradually occurring.

The main surprise during this period is that major inroads for K-12 computer science education are largely accomplished through grass-roots efforts, almost one school at a time. While a growing percentage of all school districts now offer computer science, top-down systemic buy-in for computer science at the state curriculum standards level has still not occurred in any significant way.

The CSTA has become a major force for systemic change in K-12 computer science education in the five years since it was formed. Continued progress in this area depends on the continued good work of CSTA, alongside the many grass-roots efforts of individual instructors and institutes.

Hopefully all these efforts will lead to eventual buy-in by more state education boards, and to better understanding by the general public about the nature and importance of computer science in the education of all K-12 students.

References

ACM Task Force Curriculum Committee: A Model Curriculum for K-12 Computer Science, 2nd edn. Association for Computing Machinery (2005), ISBN 59593-596-7, http://csta.acm.org/Curriculum/sub/ACMK12CSModel.html

Bureau of Labor Statistics. Occupational Outlook Handbook, 2008-2009 Edition (2009), http://www.bls.gov/oco/ocos267.htm

Goode, J.: Reprogramming College Preparatory Computer Science. Communications of the ACM 51(11), 31–33 (2008)

Salus, P.: The Daemon, the Gnu, and the Penguin: How free and open source software is changing the world. Reed Media Services (2008), Free prepublication version, http://www.groklaw.net/staticpages/index.php?page=20051013231901859

CSTA National Secondary Computer Science Survey: Comparison of 2005, 2007, and 2009 Results. Computer Science Teachers Association (2009), http://csta.acm.org/Research/sub/CSTAResearch.html

Stephenson, C.: The State of K-12 Computer Science Education. Computer Science Teachers Association Capitol Briefing (2009), http://csta.acm.org/Advocacy_Outreach/sub/CSTAPresentations.html

Perspective on Computer Science Education

Amiram Yehudai

Tel Aviv University

Abstract. I will present my personal perspective on Computer Science Education as I have seen it over the last three decades. Topics will include the interaction between research and education, the relation between College level curriculum and K-12 curriculum, and the role of programming languages and their influence on teaching. I will also discuss some aspects of the computer science curriculum for high-schools in Israel, which is undergoing revision.

J. Hromkovič, R. Královič, and J. Vahrenhold (Eds.): ISSEP 2010, LNCS 5941, p. 35, 2010.
© Springer-Verlag Berlin Heidelberg 2010

Didactics of Introduction to Computer Science in High School

Michal Armoni[1], Tamar Benaya[2], David Ginat[3], and Ela Zur[2]

[1] Weizmann Institute of Science, Department of Science Teaching,
POB 26, Rehovot 76100, Israel
michal.armoni@weizmann.ac.il
[2] The Open University of Israel, Computer Science Department
108 Ravutzky st. Raanana, Israel 43107
{tamar,ela}@openu.ac.il
[3] Tel-Aviv University, Science Education Department, Tel-Aviv 69978
ginat@post.tau.ac.il

Abstract. We present a didactical approach to the introductory computer science course in high school, and display a primary study of teachers' attitudes towards this approach. Our focus is on the presentation of computational elements and algorithm/program design, in a textbook that "zips" both theoretical and practical notions, while aiming for ease of comprehension on one hand and the development of a scientific discipline on the other. The teachers' responses to the presented approach reflect positive and constructive attitudes.

Keywords: Introduction to computer science, Didactics.

1 Background

In the last few decades, there has been a considerable amount of effort in developing the computer science (CS) curricula, first in the university level (e.g., [1, 2, 3]), and later-on in the high-school level (e.g., [4, 5, 6]). The goal of the K-12 designers in the US and Israel was to create a curriculum that could be widely disseminated and delivered to high-school students. The primary objective of this curriculum [5, 6] is that every CS student will learn and understand the nature of the field and the place of CS in the modern world. One particular principle underlying the curriculum is the interleaving of theoretical principles with application skills. This interleaving notion is specifically termed in the Israeli curriculum as the "zipper approach" [6].

The US framework, put together by the ACM K-12 Task Force and published in 2003, suggests four courses corresponding to different levels of progression in the CS studies. In Israel, the high-school CS curriculum is divided into two levels, which correspond to the common basic & advanced levels of topic studied in Israeli high schools. The basic level involves 3 units of studies, and the advanced level involves 5 units of studies. Each unit includes 90 school hours.

Israel's CS high-school curriculum was designed by a committee appointed by the Ministry of Education. The design was conducted between 1990 and 1993, and first

J. Hromkovič, R. Královič, and J. Vahrenhold (Eds.): ISSEP 2010, LNCS 5941, pp. 36–48, 2010.

implemented, using corresponding learning materials, in 1995. A detailed description of the program is given in [6, 7], along with the principles that guided its designers.

We briefly mention some of these principles that are relevant in this study: CS should be taught as a scientific field, just like Physics, Chemistry or Biology; and the program should concentrate on its key concepts and foundations. Specifically, the notion of an algorithmic problem and its solution should be emphasized. Conceptual and implementation issues should be interwoven throughout the taught materials; therefore the taught concepts will also be demonstrated, applied, and practiced through programming.

The curriculum has been updated to some extent since the first implementation, shifting for example from procedural languages (Pascal and C) to Object Oriented languages (Java and C#), and is now undergoing some further modification.

The 3-unit level of the curriculum includes the following units (90 hours each):

- Foundations of Computer Science 1
- Foundations of Computer Science 2
- A Practical Lab unit, to be selected from the following: Information Systems, Logic Programming, Functional Programming, Computer Graphics, Computer Organization and Assembly Language, Introduction to Web Programming.

The 5-unit level includes the 3 units described above, plus the following two units:

- Software Design (including primary Data Structures)
- An advanced unit, to be selected from the following: Computational Models, Operation Research, Computer Systems and Assembly Language, further Object Oriented Programming.

We, the authors of this paper, were involved in the development of the learning and teaching (teachers guide) materials of the Foundations course that comprises the first two units – Foundations of Computer Science 1 and 2 – which are mandatory units that constitute the basis of both the 3-unit and the 5-unit levels. The first two units are generally taught in 10th or 11th grade (age 16-17).

Our objective in this paper is to elaborate on the pedagogy and didactic approach which we followed and applied in the development of the course teaching materials. In addition, we display a preliminary study of CS teachers' attitudes towards the didactic approach of the teaching materials.

Previous studies focused primarily on the terms, notions, and order of the taught materials of the foundations course [8, 9], with some exceptions that elaborated on material development guidelines (e.g., [10]). The didactic approach presented here, as well as the teachers' attitudes study, focus on the core elements of computational-constructs presentation and algorithm-design assimilation.

We display our didactic approach in the next section. In the section that follows, we present results from our preliminary study with teachers. We conclude with a discussion of lessons from the study in light of the displayed didactic approach.

2   Foundations of Computer Science (1 and 2)

The course Foundations of Computer Science, units 1 and 2 (which we will denote by FCS1 and FCS2) are taught in Java or C#, according to the CS teachers' preference in each school. No preliminary knowledge is assumed. Each unit is designed for 90 school hours (3 weekly hours). According to the recommendations of the Israeli CS curriculum committee, the first unit is intended for grade 10, and the second – for grade 11; though in many schools both units are taught in grade 10.

2.1 Topics

The topics of the two course units "zip" conceptual notions with actual algorithmic (computer program) structures. In addition, they embed both algorithm design notions and object oriented elements. The former is presented in both units, whereas the latter is primarily displayed in the 2nd unit.

<u>Foundations of Computer Science 1</u>
- Chapter 1 - Introduction
- Chapter 2 - Algorithmic Problem Solving
- Chapter 3 - Basic Computation Model
- Chapter 4 - Algorithm Design Stages
- Chapter 5 - Conditional Execution
- Chapter 6 - Correctness of Algorithms
- Chapter 7 - Repetitive Execution
- Chapter 8 - Efficiency of Algorithms

<u>Foundations of Computer Science 2</u>
- Chapter 9 - The String class
- Chapter 10 - Arrays
- Chapter 11 - Classes and Objects
- Chapter 12 - Algorithmic Patterns (basic algorithmic schemes for counting, searching, sorting)
- Chapter 13 - Problem Solving

One may notice that the 1st unit includes the basic computational elements of: variables (chapter 3), conditional execution (ch 5), and repetition (ch 7), interleaved with the conceptual notions of: design stages (ch 4), correctness (ch 6), and efficiency (ch 8). The 2nd unit interleaves the essential OOP notions of objects and classes (ch 11) with the mutual algorithmic/OOP notion of patterns (ch 12), and the general notion of problem solving (ch 13).

The last two chapters, of algorithmic patterns and problem solving, display a rather novel approach for the Introductory CS course. We decided to embed them and their underlying notions in this way, based on positive findings in Muller's study [11, 12] of the benefits of teaching algorithmic design with an explicit focus on these terms [13], rather than implicit indications during the teaching of repetition.

The topics order in the above units shows that fundamental algorithmic aspects are displayed first, and object oriented notions are reached only later in the course. This "Algorithmics First, Objects Second" order is not typical, in most CS introductory courses. Yet, we believe that it embodies an order which is simpler for high-school students to follow and assimilate.

The following section presents the didactic approach in our course material development, which was supported and supervised by the ministry of Education. The course materials were developed as two FCS textbooks by the CS group of the Science Education department of Tel-Aviv University [14]. These textbooks are based on an earlier version, developed by the latter two authors and additional ones, for the procedural paradigm (without objects and patterns) in the CS group of the Science Teaching department of the Weizmann Institute of Science [15, 16]. Our textbooks appear in both a hard copy format and on the web [14].

2.2 Didactic Approach

The uniqueness of the developed FCS textbooks stems from a combination of several elements, including: the "zipper approach", the "object second" approach, the explicit embedding of algorithmic patterns and problem solving notions, and the employment of a couple of didactical principles – utilizing motivating examples, and demonstrating gradual design processes. In the previous section we referred to the first three elements. In this section we elaborate on the couple of didactic principles.

Two important objectives in our textbook writing were the ease of comprehension and the development of a scientific discipline. In order to achieve these objectives, we motivated each newly taught element, and repeatedly exemplified ordered program designs. This didactic approach, together with the web availability of the materials, was also aimed at assisting one in self-studying the materials.

The presentation of every new notion or computational element in the textbooks is motivated by an algorithmic problem (task). A topic is motivated by several algorithmic problems, each emphasizing a different element of the topic. For instance, the chapter presenting conditional execution is taught using nine different algorithmic problems, corresponding to various facets of alternations, such as: the if-then structure, the if-then-else structure, nested if structures, simple and structured conditions, and more. Each algorithmic problem is preceded by a motivating verbal introduction followed by the problem description. The motivating aspect yields the need to look for an algorithmic construct that is suitable for solving the problem.

A gradual solution is introduced to all the problems solved in the textbooks, structured in the following six steps:

i.   Analyzing several representative input values and discussing the relationship between the input and the output, as defined in the problem specification.
ii. Dividing the problem into sub-problems, or sub-tasks.
iii. Selecting appropriate data structures, which in this introductory level are mostly variables. The description of each variable includes its role, its corresponding name, and its appropriate data type.
iv. Writing the algorithmic solution in pseudo-code.
v.   Translating the algorithm into code (Java or C#).
vi. Walking through the code with representative inputs.

These stages orderly display a program design process. Experienced programmers sometimes combine, or skip one or more of these steps; yet we believe that a novice should clearly realize and practice each of them, particularly the first and the last ones. In what follows, we present two examples, of two such problems and solutions. The first problem is displayed in the FCS1 textbook for motivating if-then-else, and the second problem is displayed in the FCS2 textbook for motivating the need for arrays.

Problem 1

Problem goal: Motivating the alternation statement of `if-then-else`.

Problem statement:

> Develop an algorithm, for which the input is a three-digit positive integer (greater-than 99) and the output is a message indicating whether the integer is a palindrome.

Analyzing the input-output relationship: Examples of three-digit positive integers which are palindromes are: 787, 424 and 777. Examples of three-digit positive integers which are not palindromes are: 778, 234 and 192. The examples demonstrate that the computation needs to distinguish between two cases: one in which a given three-digit integer is a palindrome (when the *units* digit is equal to the *hundreds* digit), and one in which it is not.

Dividing the problem into sub-problems: The problem can be divided into four sub-problems:

1. Input a three-digit positive integer.
2. Extract the *unit's* digit.
3. Extract the *hundred's* digit.
4. Compare the digits; if they are equal – output a message indicating that the input integer is a palindrome. Otherwise – output a message indicating not a palindrome.

Selecting variables, including their roles, names and data types:

num – an integer variable, for storing the input.
units – an integer variable, for storing the units digit.
hundreds – an integer variable, for storing the hundreds digit.

Writing the algorithm in pseudo-code:

1. Read the input integer into **num**.
2. Assign to **units** the remainder of **num** divided by 10.
3. Assign to **hundreds** the integer result of **num** divided by 100.
4. If the value of **units** equals the value of **hundreds**
    4.1 Output the message: **num** "is a palindrome"
5. else
    5.1 Output the message: **num** "is not a palindrome"

Translating the algorithm into code (Java):
```java
/* Due to lack of space we removed the documentation */
import java.util.Scanner;
public class Palindrome
{
        public static void main (String [] args)
        {
          Scanner in = new Scanner(System.in);
          int num;           // three-digit positive number
          int units;         // unit's digit
          int hundreds;      // hundred's digit
```

```
1.        System.out.print("Enter a 3 digit number: ");
2.        num = in.nextInt();
3.        units = num % 10;
4.        hundreds = num / 100;
5.        if (units == hundreds)
              // the number is a palindrome
5.1.          System.out.println(num + " is a palindrome");
6.        else
6.1.          System.out.println(num + " is not a
              palindrome");
      }
}
```

Walking through the program: The following table presents a walk-through the program's execution, for the case of the then execution:

Table 1. Program walk-through for the input value 363

Line num	Statement	num	units	hundreds	Units ==hundreds	output
1	System.out.print("Enter a 3 digits num: ");	?	?	?		Enter a 3 Digit num:
2	num = in.nextInt();	363	?	?		
3	units = num % 10;	363	3	?		
4	Hundreds = num / 100;	363	3	3		
5	if (units == hundreds)	363	3	3	true	
5.1	System.out.println(num + " is a palindrome");	363	3	3		363 is a palindrome

Problem 2

Problem goal: Motivating the notion and utilization of a one-dimensional array.

Problem statement:

> Develop an algorithm for which the input is the running times (in seconds) of fifteen contestants in a race. The output is the number of contestants whose running times are lower than the average time.

Analyzing the input- output relationship: Given the following running times:

13 17 12 30 20 23 18 19 25 26 14 17 12 15 28

How many contestants' running time is below the average running time?

In order to solve the problem, one has to first sum all the running times, then compute the average, and finally compare each running time to the average in order to determine whether it is lower than the average.

- The sum of the fifteen running times is: 289
- The average time is: 19.27
- The number of running times lower than the average is: 9

Dividing the problem into sub-problems: In a previous chapter, we computed an average by accumulating the input values and dividing the result by the number of values. This algorithm did not store the input values in memory. But the above analysis demonstrates that <u>storing the input values in memory is necessary</u>, since they are needed for the third computation part, in which each running time is compared to the average. The problem can be divided into four sub-problems:

1. Read, store, and accumulate the running times.
2. Compute the average.
3. Count the number of running times which are lower than the average.
4. Output the number of contestants whose running times are lower than the average.

Selecting variables, including their roles, names and data type: <u>The running times are all similar in nature, and may be viewed as a sequence of integers of the same type</u>. We may identify the first value, the second value, and so on, up to the fifteenth value. Thus, we may store them all in an array.

times – an array of fifteen integers, for storing the running times.
sumOfTimes – an integer variable, for storing the total running time.
averageTime – a real variable, for the average running time.
belowAverageCounter – an integer variable, for counting the number of the below-average running times.

Writing the algorithm in pseudo-code:

1. Initialize **sumOfTimes** to 0.
2. For every i from 0 to 14 {the number of running times minus 1} do:
    2.1 Read the next running time and store it in the ith position of the **times** array.
    2.2 Accumulate the ith running time in **sumOfTimes**.
3. Divide the value in **sumOfTimes** by the number of running times and store it in **averageTime.**
4. Initialize **belowAverageCounter** to 0.
5. For every i from 0 to 14 {the number of running times minus 1} do:
    5.1 If the value of the ith running time in the array **times** is less than the **averageTime**
        5.1.1 Increment **belowAverageCounter** by 1.
6. Output the value of **belowAverageCounter.**

Translating the algorithm into code and walking through the program: The code and program walk-through are presented in the textbook FCS2 [14].

3   Teachers' Use and Attitudes towards the Didactic Approach

We conducted a preliminary study in order to examine how the didactic approach presented above is implemented by the teachers in the classroom and what are the teachers' attitudes towards it. We first briefly describe the study's methodology in section 3.1, and then display the results in section 3.2.

3.1 Methodology

Tools. We designed a nine-question questionnaire that was aimed at examining teachers' implementation of the textbooks' didactic approach. The questionnaire questions focused on the teachers' utilization of the textbooks, and their views of the textbooks as suitable for self-study.

Population. The questionnaire was sent to 24 high-school teachers who taught FCS1 and FCS2 for at least two years. Twelve teachers responded to the questionnaire.

3.2 Results

We present below a summary of the teachers' answers to the questionnaire:

Do you teach according to the materials presented in the text books?

- Five teachers indicated that they fully use the textbooks. Some comments regarding the reasons for using the textbooks were: "The books are well structured and have internal logic running through the chapters" and "The difficulty level is gradual and builds up as the content progresses".
- Seven teachers indicated that they make partial use of the textbooks. Some of the reasons given for the partial use were: "I have accumulated my own material which I am using"; "I supplement the material by using examples from additional textbooks"; "The textbooks are missing some of the required material"; "The textbooks are too wordy"; "I like to use a variety of sources" and "I also use another textbook where the problems are presented according to the level of complexity and the material is presented more graphically".

Do your students use the text books?

- Most of the teachers (10 out of 12) indicated that the students use the textbooks. The teachers described the following use by the students: "They use the textbooks as review material while preparing for the exam"; "They use the questions from the textbooks"; "They use the textbooks in order to deepen their understanding of the material presented in class" and "I actively use the textbooks in class and the students follow the material during the lesson".
- One teacher said that the students buy other textbooks but use the electronic version of our textbooks when they need a more detailed explanation.
- One teacher indicated that she hands out material in class and the students can use the textbooks in the school library.

Are the textbooks suitable for self study?

- Two teachers thought that the textbooks are fully suitable for self study. The reasons stated were: "that the book contains many solved examples starting from the design stage and ending in the implementation"; "the explanations are clear and each problem concentrates on a specific topic".
- Two teachers thought that the textbooks are not suitable for self study.

- Eight teachers thought that the textbooks are partially suitable for self study. Some of the comments made regarding the partial suitability were: "The explanations are cumbersome and do not emphasize the principle points"; "There are not enough simple examples"; "Sometimes the explanations are not elaborated enough" and "The textbooks are especially suitable for the more advanced students".

Do you refer your students to the textbooks for self study?

Most teachers (11 out of 12) indicated that they sometimes refer the students to the textbooks for self study. Some of the comments were: "I teach the basics in class and then the students use the textbooks at home in order to deepen their understanding of the material"; "I refer the students to the textbooks as a review for the material studied in class"; "I assign the material to be studied from the textbook before I present it in class"; "Students who were absent from class are instructed to use the textbooks to make up the material on their own"; and, "Before I assign a topic for self study, I highlight the important sections in the text".

Are the problems in the textbooks varied, interesting and colorful?

- Seven teachers thought that the questions in the textbooks are varied, interesting and colourful. One teacher said that although the problems are varied and interesting, the textbooks are lacking programming problems which simplify and demonstrate the material.
- Five teachers thought that only some of the problems are varied, interesting and colourful. Two teachers said that there are not enough simple problems. One teacher said that there are not enough complex and interesting problems.

Do you teach the material according to the order presented in the textbooks?

- Seven teachers indicated that they teach the material according to the order presented in the textbooks. Some of the teachers' comments were: "The material is presented in a gradual manner and logical order" and "I like the way the textbooks present the use of existing classes before the development of user defined classes."
- Five teachers indicated that they slightly change the order by either presenting static methods at an earlier stage or by implementing an object first approach.

Every topic in the textbooks is presented with a motivating example. Do you think that this is a good way to present the different topics?

- Eight teachers thought that motivational examples are a good way to present the material according to the order presented in the textbooks. One teacher said: "The student can associate the topics with the examples and thus she will remember the associative connection which facilitates the learning process". Another teacher thought: "that even if the students understand the syntax of a certain language construct, the problems clarify the motivation for this construct".
- Four teachers said that they do not always use the motivational examples. One teacher thought that the motivational examples are usually suitable for average and weak students while better students tend to skip these examples. Another teacher argued that in some of the cases the explanations are too elaborated and she thinks that it is not necessary to write out the algorithm (but only the code).

Do you present the motivating examples in class?

- Two teachers reported that they use the motivational examples in class.
- Ten teachers reported that they do not always use the motivational examples. The reasons they mentioned were: "the problems are usually suitable for average and weak students while the better students tend to skip these examples"; "sometimes I present better problems which I have accumulated throughout the years"; "I use the problems especially in order to present non trivial topics" and "I present different problems in class and I leave the motivational problems for self study".

The solutions of the motivating examples are presented in six steps. Do you present the problem solutions with these steps?

i. Analyzing several input values and observing the input-output relationship.

Table 2. Step i

Answer	N	Comments
Yes	9	"Helps to understand the problem"
No	0	
Sometimes	3	"Sometimes I analyze input values after writing the program"

ii. Dividing the problem into sub-problems.

Table 3. Step ii

Answer	N	Comments
Yes	4	
No	0	
Sometimes	8	"Not always necessary"; "The division is artificial when there is only one sub- problem"; "Sometimes the solution is intuitive"

iii. Selection of variables including roles, names and data types.

Table 4. Step iii

Answer	N	Comments
Yes	6	"Organizes the solution"
No	1	"The variables are selected as I progress with the solution"
Sometimes	5	"The variables are selected in the coding stage"; "Only when the problem is complex"; "The variables are selected in the prev stage"

iv. Writing the algorithmic solution in pseudo-code.

Table 5. Step iv

Answer	N	Comments
Yes	5	
No	1	"I discuss the algorithm in the next stage while writing the program"
Sometimes	6	"Not needed for simple problems"; "Sometimes I use a diagram"

v.   Translating the algorithm into code (Java or C#).

Table 6. Step v

Answer	N	Comments
Yes	12	
No	0	
Sometimes	0	

vi.   Walking through the code with representative inputs.

Table 7. Step vi

Answer	N	Comments
Yes	5	"It is a must"
No	0	
Sometimes	7	"I use a shorter version of walk-through"; "It is not necessary for simple problems"; "There is not always enough time in class so the students complete it at home"; "I use the walk-through especially when I teach a new topic"

4 Discussion

In examining the study results, we notice that all the teachers use the textbooks, some more extensively, and others less extensively (sometimes as supplementary materials). Some are very fond of the textbooks problems, and some see certain problems as too simple for the better students, but valuable for the less competent students. Some teachers utilize their own (collected throughout the years) materials, or more attractive graphical materials.

The majority of the teachers teach the textbooks topics in the order presented, of "Objects Second", while possibly elaborating earlier on static methods. It seems that some follow the textbook order exactly as is, while others modify, or only partially turn to some of the textbooks chapters.

The vast majority is fond of our didactical approach. They display in class some, or many of the motivating examples, or send their students to read them by themselves. And, most of them display and require algorithm design in steps, which correspond to the six steps, displayed in the textbooks materials. More than half of the teachers indicated that there is a good variety of colourful and interesting problems. The others would like to have additional problems.

Most of the teachers thought that the textbooks are only partially suitable for self study. They indeed refer their students to learn some sub-topics on their own, for preparing a new topic to be presented in class, or for deepening their understanding of a just-presented topic. It seems that the easy access to the materials on the web considerably assists teachers in referring their students to particular places in the textbooks, before or after class.

Some teachers see parts of the textbooks' text and the six code-design steps as cumbersome. This perspective is not surprising to us, as we are aware of the tendency of some programmers and teachers to skip steps and reach the coding stage as quickly as possible. We see this tendency as undesired for beginners. It seems that most of the teachers lean towards our approach, and appreciate the detailed and gradual design presentations.

All in all, we primarily focused in this paper on the didactical approach employed in our FCS textbooks for high-school CS beginners. The conducted study reveals that this approach seems beneficial and appealing to teachers. In a further study, it may be relevant to examine teachers' use and attitudes of algorithmic patterns, and their pros and cons of the "Objects Second" approach in high school.

References

1. ACM Curriculum Committee on Computer Science: Curriculum 1968, Recommendations for Academic Programs in Computer Science. Commun. ACM 11(3), 151–197 (1968)
2. Tucker, A., et al.: Computing Curricula 1991, A Summary of the ACM/IEEE-CS Joint curriculum Task Force Report. Commun. ACM 34(6), 69–84 (1991)
3. Joint IEEE Computing Society/ACM Task Force on Computing Curricula: Computing Curricula 2001 Final Report (2001),
   http://www.acm.org/education/curric_vols/cc2001.pdf
4. Merritt, S., et al.: ACM Model High School Computer Science Curriculum. Commun. ACM 36(5), 87–90 (1993)
5. Tucker, A., et al.: A Model Curriculum for K–12 Computer Science: Final Report of the ACM K-12 Task Force Curriculum Committee (2003),
   http://csta.acm.org/Curriculum/sub/k12final1022.pdf
6. Gal-Ezer, J., et al.: A High School Program in Computer Science. Computer 28(10), 73–80 (1995)
7. Gal-Ezer, J., Harel, D.: Curriculum and Course Syllabi for a High-School CS Program. Computer Science Education 9(2), 114–147 (1999)
8. Ragonis, N.: Computing pre-university: Secondary level computing curricula. In: Wah, B.W. (ed.) Wiley Encyclopedia of Computer Science and Engineering, pp. 632–648. John Wiley & Sons, Inc., Chichester (2009)
9. CSTA Curriculum Improvement Task Force: The New Educational Imperative: Improving High School Computer Science Education (2007),
   http://www.csta.acm.org/Communications/sub/
   DocsPresentationFiles/White_Paper07_06.pdf
10. Freiermuth, K., Hromkovič, J., Steffen, B.: Creating and Testing Textbooks for Secondary Schools. In: Mittermeir, R.T., Sysło, M.M. (eds.) ISSEP 2008. LNCS, vol. 5090, pp. 216–228. Springer, Heidelberg (2008)
11. Muller, O.: The effect of pattern-oriented instruction in computer science on algorithmic problem-solving skills, Ph.D. dissertation (in Hebrew), Tel-Aviv University (2007)
12. Muller, O.: Pattern Oriented Instruction and the Enhancement of Analogical Reasoning. In: Proceedings of the 1st ICER Workshop, pp. 57–67 (2005)
13. Ginat, D., Haberman, B., Cohen, D., Katz, D., Muller, O., Menashe, E.: Design Patterns for Fundamentals of Computer Science (a Hebrew textbook). Tel-Aviv University (2001)

14. Benaya, T., Armoni, M., Bilczyk (Soffrin), Y., Gradovitch, N., Green, A., Menashe, E.: Fundamentals of Computer Science in Java / C#, 2nd edn., vol. 1, 2, (Ginat, D., advisor), Tel-Aviv University, Haifa, Israel: Hashraa (in Hebrew) and electronically (2007), http://www.tau.ac.il/~csedu/yesodot.html
15. Ginat, D.: Fundamentals of Computer Science 1. Weizmann Institute of Science, Science Teaching Dept, Rehovot, Israel (1996) (in Hebrew)
16. Ben-Ari, M., Lichtenstein, O., Machlev, H., Reich, N.: Fundamentals of Computer Science 2. Weizmann Institute of Science, Science Teaching Dept., Rehovot, Israel (1998) (in Hebrew)

Software Design Course for Leading CS In-Service Teachers

Ofra Brandes[1], Tamar Vilner[2], and Ela Zur[2]

[1] The Hebrew University of Jerusalem, Science Teaching Center, Edmond J. Safra campus,
Givat Ram, Jerusalem 91-904, Israel
brandes@huji.ac.il
[2] The Open University of Israel, Computer Science Department,
108 Ravutzky St. Raanana, Israel 43107
{tami,ela}@openu.ac.il

Abstract. The Computer Science (CS) discipline is continually developing. Consequently there are frequent changes in curricula and their implementations. The CS teachers, who usually work in schools alone or in small teams, are seriously challenged by the changes, since all their teaching materials and pedagogical methods have to be revised. In Israel, the secondary school CS program has recently shifted from the procedural to the OOP paradigm. This paper discusses an approach taken to address the ensuing difficulties, namely a course for leading teachers. The paper describes the rationale of leading teachers, and a course conducted for them in an advanced unit in the CS program. The main purpose of the course was to develop a professional leadership of CS teachers who can support and contribute to their peers. The paper describes the expectations of the course; its progress; and the conclusions about its success obtained by analyzing data collected during the course.

Keywords: In-Service Teachers, Leading Teachers, Software Design.

1 Introduction

1.1 The High School CS Curriculum

During the last four decades there has been considerable activity surrounding CS curricula at all levels. The rapid changes which have taken place in computer science and in the computer industry entailed frequent changes of curricula, as can be seen in the series of curricula which were developed during those years [1, 2, 3]. These changes had, in particular, a direct effect on the high-school curriculum [4, 5]. Today there are many different approaches to CS teaching in secondary school [6].

Israel's high school CS curricula was designed between 1990 and 1993 and first implemented in 1995 based on the procedural approach (Pascal or C). A detailed description of the program and the principles that guided the work of its designers is given in [7, 8]. The curriculum contains individual units, each of 90 hours. Two programs are offered, consisting respectively of three and five units. The 3-unit program includes two mandatory core units, Fundamentals 1 and 2, which present the foundations of

J. Hromkovič, R. Královič, and J. Vahrenhold (Eds.): ISSEP 2010, LNCS 5941, pp. 49–60, 2010.

algorithmic thinking and programming. The 5-unit program is intended for more advanced students. It includes the 3-unit version and the fourth mandatory unit, Software Design, which is an extension of Fundamentals 1 and 2. The third and the fifth units can be chosen from several alternatives.

Following the shift in the programming world from procedural to object-oriented programming (OOP), the three units, Fundamentals 1 and 2 and Software Design, were re-implemented in the OOP paradigm. This paper deals with one aspect of this change, relating to the software design unit.

1.2   Object Based Software Design Unit

The software design unit concentrated on data structures, and abstract data-types. It also took a step beyond stand-alone algorithms, to discuss the design of complete (small) systems [8]. A few years ago, about ten years after the development of the unit was completed, the unit was rewritten using the OOP approach [9]. It can now be taught in Java or in C#. Accordingly, it is now called Object Based Software Design (henceforth "the unit").

The unit's goals, which remained almost the same despite the change in paradigm, are as follows: (a) to present the principles of a systematic approach to programming, especially the need and the benefit of dividing a problem into sub-problems; (b) to teach basic concepts of complexity illustrated by discussions of various algorithms; (c) to develop abstract thinking, especially the separation between specification (and the interface that reflects it) and implementation; (d) to teach how to define data structures and abstract data types, pointing out the difference between them. The last two make up the bulk of the course. In the procedural approach they were implemented by using the "unit" concept of Turbo Pascal. In the new OOP approach classes are used instead.

The content of the revised unit is as follows [10]:

• An introduction to Objects and Classes • Recursion • Complexity • Objects and Classes revisited • Data collections • Stacks and Queues • List – a linear collection • Binary Tree – an hierarchic collection • Map – mapping keys to values – a matching collection.

Although the unit's goals remained largely unchanged, the revision was not a simple translation. The differences between the paradigms forced many subtle changes in content and presentation of the unit. Thus, the in-service CS teachers were confronted with the need to master a new paradigm, to rewrite their teaching materials and to make an effort to understand the revised unit

1.3   Object Based Software Design Course for Leading Teachers

Machshava is the Israeli National Center of High School Computer Science Teachers [11]. One of its main goals is to foster a professional leadership of CS teachers, in accordance with the belief that an organized group of leading teachers (LTs) can serve as a model for other teachers, promote pedagogical objectives, inspire other colleagues and help them adjust to new courses and topics [12, 13, 14]. In order to foster such leading groups, *Machshava* organizes LTs courses. An LTs course has three main goals: to enhance the participants' understanding of new topics and materials; to strengthen the group as a professional community; and to foster leadership growth [13].

In 2008, we, the unit development team (henceforth "the team"), adopted this model and conducted a course for in-service LTs, dealing with the Object Based Software Design unit. The 112 hour course was a mixture of monthly meetings, a three-day summer seminar, and virtual work through forums. Our goals and expectations of the course were as follows:

1. To enhance the participants' understanding of the unit contents
2. To train the participants to guide and direct regional pedagogical workshops for their peers
3. To extract and explicitly formulate the practical experience of the participants so this knowledge could be distributed to the whole CS teachers' community.

Our first goal is identical to that of *Machshava*. Our second and third goals are concrete, practical ways to achieve *Machshava*'s last two goals.

During the course, the participants were required to implement a project using the new concepts presented in the unit (cf. point 1 above); some were required to teach regional pedagogical workshops for their peers (cf. 2 above). These LTs will be referred to, as "workshop-teachers". The participants were also asked to prepare learning and teaching materials for the CS teachers' community (cf. 3 above). In addition, the workshop-teachers had to report on their experience, achievements and difficulties in their workshops. The other course participants discussed these reports and helped the workshop-teachers to address their difficulties.

Acceptance into the LTs course was based on the candidates' potential to lead and inspire other teachers [13]. Each candidate had to be an in-service CS high school teacher; to possess previous knowledge of Java or C#; to be teaching the unit during the current year; and to agree to guide a workshop. In order to establish workshops throughout the country, we were forced to accept some teachers who were not teaching the unit during the current year, because of their region of residence. In some other regions, we had more candidates than were needed.

The purpose of our research was to examine the extent to which our expectations from the course were fulfilled, and to identify factors which affected, positively or negatively, the success of the course.

2 The Research

We accompanied the LTs course by ongoing qualitative research in order to estimate the extent to which the team's goals for the course were achieved. Most of the qualitative research in computing education deals with the students' side and the subject matter [15]. Our research explored the in-service teachers' point of view following the change in the CS teaching paradigm.

2.1 Research Questions

The research questions refer to the course goals and to the team's expectations. To examine the extent to which these goals were fulfilled, we examine them one by one:

1. Did the LTs gain an enhanced understanding of the unit contents?
2. Did we succeed in training the LTs to guide and direct regional pedagogical workshops for their peers?
3. Did the course succeed in creating a professional leadership of CS teachers who feel responsible to the CS teachers' community?
4. What factors in the course structure affected, positively or negatively, the fulfilment of the team's expectations?

Research questions 1 and 2, directly check our first and second goals (cf. 1.3 above). The third research question, which expresses *Machshava's* general aims, will be answered in concrete, practical ways as described in Section 4.3. The fourth question is designated to explain the achievements of the course.

2.2 Research Population

The research population consisted of 25 teachers:

- • Seventeen teachers (68%) had a B.A. or B.Sc. in CS
- • Six teachers (24%) had a B.Sc. in other sciences
- • Only two (8%) had a B.A. in other, non-scientific, disciplines.

Seventeen teachers had a second degree (M.A. or M.Sc.) in various disciplines, such as CS, Education and others.

Figure 1 presents the teachers' previous knowledge of Java and C# as reported by the teachers in the first questionnaire (see 2.3 below):

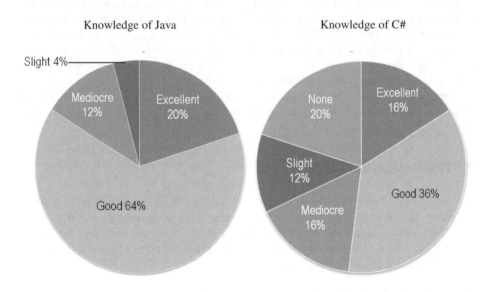

Fig. 1. Teachers' previous knowledge of Java and C#

2.3 Research Instruments

In order to answer the first three research questions, we chose to rely on the participants' narratives and to conduct qualitative-constructivist research based on Shkedi's constructivist methods of qualitative research [16].

The course participants were asked to complete a questionnaire before they started the course. The questionnaire contained questions about their background and their expectations of the course. We used the Grounded Theory methodology in order to define a categorical system which represented the teachers' expectations of the course. These expectations and our goals, as described above, serve as the basis for the research which followed. Throughout the course we wrote field notes, which helped us to record the general mood and events.

Another questionnaire was handed to the teachers at the beginning of the summer seminar, which ended the LTs course. This questionnaire focused on their attitudes and understanding of the concept of an LT. In addition, we conducted two discussions with the participants about their perceived benefits and achievements from the course, at the beginning and the end of the summer seminar. We observed and participated in those discussions, which were recorded by audiotape and later transcribed.

The field notes, the transcription of the discussions, and the answers to the two questionnaires are the data we used to answer our research questions.

3 Analysis

The answers to the first questionnaire reflect the participants' expectations of the course. Their answers to the second questionnaire and their comments during the discussions reflect their conclusions and views of the course and its benefits. In this section we analyze these data.

3.1 The First Categorical System

The first questionnaire, delivered before the course began, was answered by all the LTs. We collected 59 quotes, grouped similar quotes and then divided the LTs' expectations into three main categories: the individual aspect, the social aspect and the community aspect. These categories were divided again into sub-categories. Table 1 shows the whole categorical system. The numbers in the table represent the number of quotes written by the LTs about each category, in their 'emic' language [17] (with minor necessary changes due to the translation from Hebrew to English).

Table1 shows that 56% of the quotes (33) were centered on the **individual aspect**. This aspect expresses the team's first goal. As can be seen from the table, most of the LTs wanted the course to contribute to their own knowledge. In contrast to the teachers, the team thought of this only as an essential starting point, and not as the main goal of the course. It certainly was necessary for the LTs to understand the changes and the novelties of the new unit. But this was only the base for the main goals (cf. 2-3 in 1.3 above), of fostering a professional leadership that will contribute to the CS community.

Table 1. The first categorical system: teachers' expectations from the LTs-course

Categories	Sub-categories	Examples of quotes
Individual aspect (33)	Content enrichment (16)	• *"reinforcement and deepening in the unit material"* • *"internalizing the Object Oriented programming subject"* • *"to learn much more than the material taught in the class"*
	Extending and developing the questions pool (4)	• *"to finish the course with an extra pool of questions. To finish with additional value"* • *"to build a good pool of questions"*
	Didactic enrichment (13)	• *"another point of view about teaching the subject"* • *"enrichment of ways to teach the students"* • *"to know the new unit – what are the changes? and to try to enhance my understanding of the unit to the best level, so I will be ready to teach my students"* • *"before I get into my class, I would like to learn how and what to teach"*
Social aspect (13)	Psychological support group (2)	• *"creating a cooperation between the teachers. It is great to cooperate. At school I'm alone"*
	Content support group (11)	• *"the interaction with the unit development team will bring new ideas and will open the mind"* • *"to arouse intellectual stimulation"* • *"to learn from others' experience"* • *"to hear ideas and opinions from peers"* • *"hope to check whether the previous decisions made by our team at school are compatible with the others' opinions"*
Community aspect (13)	Contribution to the community (8)	• *"to help novice teachers at the beginning of the road... I know I'll gain as much as I will contribute"* • *"The possibility of transferring knowledge since I feel that many teachers in the field didn't internalize the contents of the old "software design" unit, and I doubt if they are able to teach the students the new contents"*
	Influencing the content of the software design unit (5)[1]	• *"to try to influence the topics to be included or changed in the future"* • *"to contribute to the improvement of the unit content"*

---

[1] These expectations could not be realized because the unit had already been specified by that time.

The **social aspect** in the categorical system could match - if directed towards the community - the idea of a professional leadership. But in fact it seems from the quotes that here too the LTs were directed towards their own benefits. And so another 22% of the quotes (13), are oriented towards the LTs self-benefit.

Contrasting with this 78%, the remaining 22% of the quotes (13) show a **community aspect,** and can be directly related to our main goals.

3.2 The Second Categorical System

While there was some correlation between the LTs' expectations and the team's goals, the emphasis was quite different. As mentioned before, at the end of the course we held a three-day summer seminar [13, 14]. Most of the LTs (19 out of 25) participated in this seminar, as well as 11 academics and other teachers who joined us. The data we collected from the second questionnaire, the discussions and the field notes enabled us to collect 65 quotes, 62 from the LTs, and three from the newcomers, which we ignored. These quotes were classified using the same categorical system (Table 2). A comparison of the two data sets was used to answer the first research question. (Quotes illustrating some of the categories will appear below, in Section 4.)

Table 2. The second categorical system: teachers' attitudes to the LTs course.

Categories	Sub-categories
Individual aspect (11)	Content enrichment (9)
	Extending and developing the questions pool (0)
	Didactic enrichment (2)
Social aspect (17)	Psychological support group (11)
	Content support group (6)
Community aspect (34)	Contribution to the community (27)
	Influencing the content of the software design unit (7)

4 Results and Discussion

It is quite clear that Table 2 expresses changes of attitude. Analyzing the results enables us to understand the changes which took place.

4.1 Enhanced Understanding of the Unit Contents

The first goal of the course was to enhance the participants' understanding of the unit contents. All the LTs were experienced teachers and knew the basic contents. In order to achieve this first goal, we exposed the LTs to the changes caused by the paradigm shift and to various issues that influenced the contents and organization of the revised unit. The teachers' feeling at the end of the course was that their knowledge was extended and enhanced, as illustrated by the following quotes.

T1, who expected at the beginning of the course to *"help myself to know the subject much better"*, said at the end, after successfully guiding a workshop for peers: *"we have been in the field for only one year, but now I feel like I have four or five years' experience."* T2 related to the start and the end points: *"As a result of the deep learning, I understood how big the change in the unit was. When I started the course, I thought I knew the material very well, but now I understand that I knew it only superficially. Without the course we wouldn't be exposed to the unit nuances."* T3, who claimed that generally she adjusts to changes very easily, said *"there are emphases in the unit that I wouldn't pay attention to, by just reading the text."* T4 said: *"After participating in these meetings, I found myself in another place than if I had not come."* Participating in the LTs course in addition to guiding the workshops, gave the participants a better understanding of the unit contents and the changes caused by the paradigm shift.

As mentioned previously, we were concerned with the fact that the majority of teachers (78%) emphasized self-benefit before starting the course. Hence we were pleased to see that at the end of the course, the situation was quite different. We counted only 45% of quotes (28) about this topic, and even then, the teachers talked more about understanding the differences and the meaning of the paradigm shift, and not about their basic knowledge of the material. This is also evident from the quotes above.

4.2 Guiding Workshops

The second goal of the course was to train the participants to guide workshops for their peers. During the LTs course and the following months, six pedagogical workshops guided by 13 of the LTs were conducted in different regions. Around 180 teachers participated in these workshops. We conducted close support and monitoring activity; constructive criticism was given to the workshop-teachers by us and by the other LTs who got progress reports from their colleagues. We observed the workshops meetings, insisting on not being involved in them. We noted issues that were not treated well or clearly enough, and during the summary meetings of the workshops we discussed and clarified those issues. It is clear that the LTs had covered most of the necessary material - the unit contents and most of the issues which influenced the unit organization - in a satisfactory way. Thus, it seems that our second goal was achieved. From various comments made by the LTs, it seems that, beside teaching their peers, the LTs accumulated a lot of experience and enhanced their own knowledge (cf. 1.3 above). All this is expressed by the following quotes:

- T5, a senior, successful teacher, expresses the self-actualization benefit: *"It was a fascinating year... maybe because we had to teach the other teachers."*
- T6 said similar things: *"Now, with the workshop we are teaching... despite the difficulties... I think it contributes a lot to those who teach. I hope that we also contribute a lot to the teachers in the field, but to us, from the point of view of enhancing our knowledge and facing questions where you really have to ask yourself: can I really answer? and do I really have an answer? and do I really understand why is it so...? It contributes a lot and people benefit from it."*

It seems that the workshop-teachers benefited more than the LTs who didn't guide workshops, though they shared the process of supporting the workshop-teachers and dealing with their doubts. This can be seen from their words: *"I didn't go through the process they did... I'm sure they learned a lot... The benefit to yourself for anyone who did it is enormous because when you are standing, especially in front of teachers, you are deliberating how to correctly teach the point."*

The idea that guiding workshops contributes directly to the knowledge, self-confidence and professional development of the workshop-teachers is important. We will return to it while discussing the last research question.

4.3 Creating CS Leadership

The third research question was: "Did the course succeed in creating a professional leadership of CS teachers who feel responsible to the CS teachers' community?" The team's second and third goals (cf. 1.3 above) are concrete, practical ways to achieve this purpose.

It is clear, as can be seen from the comparison between Table 1 and Table 2, that a change had occurred in the LTs' attitude during the course and they understood that contributing to the community is an important aspect of the course. Whereas only 14% of the quotes in Table 1 relate to the "contributing to the community" aspect, Table 2 shows 44% of quotes relating to this aspect. The following are some of the quotes that express this change:

- T7: *"We helped them to take the first steps - help which we didn't have when we started a year ago... this means to lead...we are leading someone else after us."*
- T3, who was quoted above (cf. 4.1.), continued describing her experience: *"Only after meeting with the teachers in the workshop, I realized how important what we are doing is. Moreover, the feeling of... a mission, as much as it sounds a grandiose word, became stronger, after every such meeting."*
- T2 added in the same manner: *"...I realized how big the change in the unit's content is and how much the teachers in the field are not aware of it at all. ... I met teachers who have begun [to teach the new unit] but they don't understand what they began at all, they asked such questions that I realized even more how much they are stuck in the mud. Then I understood that we must very quickly bring it to people somehow."*

The following activities of the LTs during 2008-2009 indicate the growth of the CS leadership:

1. The LTs who didn't guide workshops were asked to contribute to the community by supplying answers to an existing pool of questions and transferring an existing laboratory collection from Java to C# and vice versa etc. All these products were published on open sites for CS teachers.
2. In one workshop, the LTs voluntarily doubled the workshop length (56 hours instead of 28). In the first 28 hours they taught the material. In the additional 28 hours they worked together with the teachers on writing and creating new questions which check understanding of important issues in the new unit.

3. We conducted a closed web-forum for the LTs, in which the LTs expressed and discussed difficulties and reinforced their colleagues. At the end of the course this forum was terminated, but most of the LTs continued to be active members in the Object Based System Design web-forum, open to all CS teachers. There the LTs presented new questions and answers which they used in their classes. Towards the matriculation exams (May 2009), they also published examples of exams.

Preparing learning materials (cf. point 1 above), or participating in the course forum, may be viewed as carrying out the tasks of the course, rather than an indication of professional leadership growth. However, points 2 and 3 above can only be viewed as voluntary activities, motivated by a feeling of responsibility for the community at large. This, and the feeling about the significance of the workshops for the community discussed in Section 4.2, indicate the success of creating a professional CS leadership.

4.4 Factors Affecting the Course

The last research question was: "What factors in the course structure affected, positively or negatively, the fulfillment of the team's expectations?" Here we suggest some factors that might contribute to this success.

The experience of guiding workshops. Conducting the workshops was a positive experience for the LTs, and they pointed out that they had learned a lot from it (cf. Section 4.2.). It contributed a great deal to their own enhanced understanding and to their teaching experience. They also felt it addressed an important need of the community. They had discovered the ability to lead others and help them.

The direct connections between the LTs and the team. We attributed great importance to the direct connection with the LTs, without mediators. The LTs expressed a similar feeling about the direct contact with the team, mapped under the sub-category: "Content support group". The following quotes from the last meeting in the seminar express the importance of this issue: *"First of all you need someone professional to guide you in the context of the new material. Even if you learn alone... it's never the same as if someone comes to you and gives you his views of the new thing... I think you can't [understand] if you don't have someone from that unit development team."*

The existence of a support group among the LTs. The sub-category "Psychological support group" had only two quotes (3%) before the course, but became more significant at its end, with 11 quotes (17%). The fact that one does not have to face the revolution alone and invent everything from scratch contributed to the LTs' self-confidence. The participants learned to appreciate one another and found that they can gain from their interactions. T6 is quite a senior teacher. Though she has many friends who are CS teachers, she felt that they could not help her as much as the support group: *"To me, personally, it helped a lot because in school I am all alone, teaching this unit This was an excellent support group with regard to both the deepened understanding of the material and what I heard, also the indecision and difficulties of people. All this connected to my indecision and also the experience with the workshops."* T9 is a lonely, relatively new immigrant teacher: *"...I am the only one in school, I'm really alone and* *suddenly I felt myself belonging to a group of people from the same field."* T7 has taught in the university for many years, helps the CS supervisor in the Educational Ministry, and has good and continuing relations with many teachers, novices and professionals. Even she comments: *"Being an LT gives me confidence in the teaching process. The acquaintance with other LTs gives me the option of an additional support group to strengthen my self-confidence."*

5 Conclusions

The high school CS curriculum changes often, in line with the nature of the CS discipline. Every change frightens even senior CS teachers. All the existing teaching materials gathered over the years, become obsolete. The teachers' self-confidence is harmed, their rich pedagogical experience which helped them to identify misconceptions and comprehension difficulties is no longer pertinent. The fact that CS teachers work in schools alone or in a very small team also affects their ability to cope with innovations.

In order to at least partially address the problems just described, we conducted an experimental LTs course. This course was intended to enhance the participants' understanding of the contents of a revised CS unit, to train the participants to guide and direct pedagogical workshops for their peers, and to extract and formulate their practical experience so this knowledge could be distributed to the whole community of CS teachers.

We accompanied the LTs course with qualitative research in order to estimate the extent to which the team's goals were achieved. From Table 1 we can conclude that before the course, the teachers were mainly interested in enhancing their own knowledge, and were not sufficiently aware of the extra meaning of being an LT. The course process led them to an understanding of the need for LTs. They also realized that they can serve as the needed professional CS leadership, as can be seen in Table 2. The current results show that most of our goals were achieved. Primarily, the subjective interest in the course had changed during the process to a communal one.

The following factors can explain at least some of the achievements:

- Social interaction between the teachers themselves.
- Close interaction with the team developing the unit.
- Practical experience in guiding pedagogical workshops for LT colleagues.

This research concentrated on the LTs course itself. However, only long term followup can determine whether its effect is long-lasting, and whether the participants have indeed formed a professional leadership for the teachers' community.

Acknowledgments. We thank Dr. T. Lapidot, manager of *Machshava,* the initiator and producer of the LTs course, and all the leading teachers of the 2008 course. We also thank Prof. C. Beeri, the scientific advisor of the team who helped in the preparation of this paper, and Y. Adamovsky and I. Peretz, central members of the development team.

References

1. ACM Curriculum Committee on Computer Science: Curriculum 1968 Recommendations for Academic Programs in Computer Science. Comm. Assoc. Comput. Mach. 11, 151–197 (1968)
2. Tucker, A., et al.: Computing Curricula 1991: A Summary of the ACM/IEEE-CS Joint curriculum Task Force Report. Comm. Assoc. Comput. Mach. 34, 69–84 (1991)
3. Joint IEEE Computing Society/ACM Task Force on Computing Curricula. Computing Curricula 2001 Final Report (2001),
   http://www.acm.org/education/curric_vols/cc2001.pdf
4. Merrit, S., et al.: ACM Model High School Computer Science Curriculum. Association for Computing Machinery, New York (1994)
5. Tucker, A., et al.: A Model Curriculum for K–12 Computer Science: Final Report of the ACM K-12 Task Force Curriculum Committee (2003),
   http://csta.acm.org/Curriculum/sub/k12final1022.pdf
6. Ragonis, N.: Computing Pre-University: Secondary Level Computing Curricula. In: Wah, B.W. (ed.) Wiley Encyclopedia of Computer Science and Engineering, vol. 5(1), pp. 632–648. John Wiley & Sons, Inc., Hoboken (2009)
7. Gal-Ezer, J., et al.: A High-School Program in Computer Science. Computer 28(10), 73–80 (1995)
8. Gal-Ezer, J., Harel, D.: Curriculum and Course Syllabi for High-School Computer Science Program. Computer Science Education 9(2), 114–147 (1999)
9. Brandes, O., et al.: Software Design. The Hebrew University, Science Teaching Center (1997) (in Hebrew)
10. Brandes, O., et al.: Object Based Software Design, The Hebrew University, Science Teaching Center (2007) (in Hebrew)
11. Israel National Center for Computer Science Teachers.: "Machshava" – The Israeli National Center for High School Computer Science Teachers. In: Proceeding of the 7th SIGCSE Annual Conference on Innovation and Technology in Computer Science Education, Aarhus Denmark, p. 234 (2002)
12. Lapidot, T., Aharoni, D.: The Israeli Summer Seminars for CS Leading Teachers. In: Proceeding of the 12th SIGCSE Annual Conference on Innovation and Technology in Computer Science Education, Dundee Scotland, UK, p. 318 (2007)
13. Lapidot, T.: Supporting the Growth of CS Leading Teachers. In: Proceeding of the 12th SIGCSE Annual Conference on Innovation and Technology in Computer Science Education Dundee Scotland, UK, p. 327 (2007)
14. Lapidot, T., Aharoni, D.: On the Frontier of Computer Science: Israeli Summer Seminars. Inroads - SIGCSE Bulletin 40(4), 72–74 (2008)
15. Berlund, A., et al.: Qualitative Research Projects in Computing Education Research: An Overview. In: Proceedings of the 8th Australian conference on Computing education, Hobart Tasmania, pp. 25–33 (2006)
16. Shkedi, A.: Words of Meaning: Qualitative Research - Theory and Practice. University of Tel Aviv Press, Tel Aviv (2004)
17. Patton, M.Q.: Qualitative Evaluation and Research Methods. Sage Publications, Beverly Hill (1990)

The Effect of Tangible Artifacts, Gender and Subjective Technical Competence on Teaching Programming to Seventh Graders

Philipp Brauner[1], Thiemo Leonhardt[1], Martina Ziefle[2], and Ulrik Schroeder[1]

[1] Lehr- und Forschungsgebiet Informatik 9
RWTH Aachen University
Ahornstr. 55, 52056 Aachen, Germany
{brauner,leonhardt,schroeder}@cs.rwth-aachen.de
[2] Human Technology Centre (HumTec)
RWTH Aachen University
Theaterplatz 14
52056 Aachen, Germany
ziefle@humtec.rwth-aachen.de

Abstract. This study compares the effect of using tangible robots to using visual representations for introducing seventh graders (12 to 13 year old) to computer programming. The impact was measured on learning outcome, self-efficacy, class feedback and attitudes towards STEM (science, technology, engeneering and mathematics) topics. Results show that using robots to learn computer programming is beneficial, although no overall effect towards STEM topics could be shown. A huge gender gap in regard to subjective technical competence (STC) was found that negatively affected the participants' performance. We provide approaches to leverage this gap and increase learning outcome and interest in STEM topics.

Keywords: Gender effects, subjective technical competence, introductionary programming, CS0, Robots, LEGO Mindstorms, self-efficacy.

1 Introduction

Despite the economic crisis, demand for IT-professionals persists. In addition to the possibility of occupational retraining of professionals, the goal must be to optimize teaching at schools in quality and quantity. Therefore education in school has to improve young students' interest in technology and in particular in STEM subjects. Another aspect is the low number of female students in STEM subjects. According to studies of the German education system women more often than men choose disciplines like linguistics, cultural studies, fine arts and human or veterinary medicine [1]. Already in school girls show less interest in topics of computer science in Germany [2].

To increase the participation in STEM in general and the participation of women in particular a lot of projects have been initiated. One popular concept is to utilize robots as a tool to increase interest in programming. One of our research goals was to determine if

J. Hromkovič, R. Královič, and J. Vahrenhold (Eds.): ISSEP 2010, LNCS 5941, pp. 61–71, 2010.

the tangibility of a robot is actually necessary. In a controlled experiment we compared the effect of tangible robots to the use of visual representations of robots for introducing school students to programming. We measured their impact on learning outcome, self-efficacy, class feedback and attitudes towards STEM topics.

2 Related Work

2.1 Importance of Self-efficacy and Self confidence

Self-efficacy [3] refers to the individual confidence in one's capability to execute a certain behavior or archive a certain goal. Specifically technology self-efficacy is an effective variable as it determines not only users emotional attitudes towards own abilities, but also the open-mindedness to frequently interact with technology, as any anticipated failure when interacting with technology is avoided, which also results in a low computer experience. Studies have shown that high scores in computer self-efficacy are related to performance with and acceptance of technology [4, 5, 6]. In addition, there are profound gender differences regarding technical self-confidence. Women usually report lower levels of computer-related self-efficacy and higher computer anxiety [7], which in turn, reduces the probability of active computer interaction and may lead to a generally lower computer-expertise level. Furthermore it was found that self-efficacy is an important factor for academic decisions and the career development of women working in the STEM area [8, 9]. Self-efficacy is mainly constituted by role models and social persuasion. It is assumed that high self-efficacy is especially important for women due to their low representation in STEM.

2.2 Effect of Robot Courses

Various experiences indicate high learning motivation by girls and young women in robot courses [10]. Thus learning with and about robots, as an analogy to learning about engineering, is a well established measure. However, the learning objective in STEM classes must also be experienced as useful. This approach follows the assumption that girls reject the principle of "technology for technology's sake" and, in contrast to boys, do not regard the "feasible" as the "useful" [11].

3 Method

To evaluate whether programming tangible artifacts like robots is advantageous over programming virtual agents displayed solely on a computer screen, a controlled experiment in form of a school lesson was carried out. The modality of the robot was a between-subject variable, i.e. one group of students interacted with a tangible robot in form of a LEGO Mindstorms NXT and one group worked with a visual image of a robot presented solely on the computer screen.

This section describes the targeted audience and the composition of our groups, the teaching unit and the programming language used in this experiment, as well as the experimental setup and the variables assessed.

3.1 Targeted Audience and Group Composition

The experiment was carried out with seventh graders (12 to 13 years old), because the projects *"Mädchen machen Informatik"* ("girls do informatics") at TU Munich[1] and *"go4IT!"* at RWTH Aachen University[2] both successfully concentrate their STEM-support-projects on students at that age. The projects try to achieve sustainable effects with their interventions at important points in the students' biography, e.g. when seventh graders select their specialization courses for the eighth grade. Here, foreign languages compete with STEM courses and an early decision for the students' future career is made.

According to recommendations of the Roberta project[3] in courses the student to tutor ratio should not exceed eight to ten pupils per tutor. The projects *"go4IT!"* and *"Mädchen machen Informatik"* choose a lower supervision ratio of six pupils per tutor. Also they suggest learning in pairs. This approach is based on the recognition that girls and young women use computers cooperatively and utility-oriented [12].

The groups were mixed-gender and thereby followed the latest concepts of gender sensitive workshop design [13]. This is based on the finding of the research project "Roberta", that no gender effects are expected in coeducational courses as long as one follows a gender sensitive course design. We followed recommendations like promoting interest, self-awareness and self-confidence, recognition of learning achievement, activating of social competences and avoid interferences [10].

3.2 Programming Environment

The programming language Scratch [14] was used in this study, as syntax errors are avoided by arranging command blocks by mouse and textual commands need not be remembered as they can be looked. A simple robot program is "written" by connecting movement blocks with rotation commands. Adding blocks for loops or conditions allow the development of more complex programs.

In addition, Scratch innately supports the turtle metaphor we based our teaching unit on. Originally the visual turtles on Scratch's stage are controlled by movement and sensory commands. These commands were modified in a way that they could be linked with a LEGO Mindstorms NXT robot over a Bluetooth connection. To make sure that both experimental conditions perform similarly aside from the different representation of the robot, we made sure that the same delays were present in both conditions. In addition the sprite of the robot was hidden for the group with the actual robot to avoid presenting them the robot twice.

3.3 Teaching Unit

We developed a teaching unit of 90 minutes that each group had to undergo. The course started with an introduction of ourselves and assured that we're not interested in rating the students but in finding better approaches for teaching introductory programming. The students filled out the first questionnaire (see below) and learned how

---

[1] http://portal.mytum.de/am/mmi/
[2] http://lehramt.informatik.rwth-aachen.de/go4it
[3] http://www.roberta-home.de/de

the robot can be programmed by connecting Scratch's command blocks (15 min.). The robot was programmed using "turtle talk" [15], i.e. by issuing relative movement commands like forward, turn left 90°, etc. This programming concept is easy to understand by students without prior programming knowledge.

The first task was writing a computer program that let the turtle follow simple geometric patterns by using only movement commands (20 min.). In the second task more complex geometric figures had to be programmed which made the introduction of loops necessary (20 min.). After the first two tasks the final questionnaire was filled out. To let every student get in touch with a LEGO Mindstorms robot, the third assignment was to write a robot program that used the light sensor as input for preventing the robot to fall off the table. This task was carried out after the final questionnaires, therefore it had no impact on the data presented in the results section.

3.4 Experimental Setup

The experiment was carried out with 31 children from a 7th grade of a local school. The children were 13 or 14 years old; 16 of them were male, 15 were female. Due to sickness or shifts in the schools time tables we were not able to balance the groups perfectly: Thus the first group consisted of four girls and four boys, the second of five girls and four boys, the third of four girls and three boys, and the last of three girls and four boys. The first two groups worked with the tangible robot, the third and forth group worked with the visual turtle presented on the computer screen.

Each group of students was separated into teams of two with each team sitting in front of one of four tables. The tables were evenly arranged around two center tables that formed a shared space for testing the robots' actions. Each team had a pre-constructed LEGO Mindstorms NXT robot and a laptop with Scratch running in full screen mode. Changing window size or switching tasks was disabled.

3.5 Variables

The participants' sex and the modality of the turtle were used as independent variables. As depended variables we assessed subjective ratings of the students' competencies, their weekly computer usage, learning outcome, and class feedback.

3.5.1 Learning Outcome

For measuring learning outcomes the students had to work on two tasks. The first was understanding a given Scratch program by drawing the path of the robot. The second task was writing a program that makes the robot move along a given path. As we wanted to assess the students' understanding of computer programming concepts and not their ability to remember commands, all command blocks necessary were printed on the exercise sheet, as well as some additional ones as dummies.

For each of the tasks a list of five criteria was defined (e.g. correct number of loop repetitions, turning the robot in the right direction). For each criterion present in the students' solution a point was given. Therefore in each task the students could earn zero (no or completely wrong solution) to five points (correct solution). Much effort was put into the grading as some students did not use the puzzle syntax for writing down their code. One student consecutively numbered the pieces and wrote down a (correct) sequence of numbers instead. Hence his solution was graded five points.

3.5.2 Class Feedback

We assessed four factors as class feedback whereby each factor consisted of multiple questions: The liking of the class (3 items such as "I found the class interesting"), the perceived simplicity of the class (3 items such as "I understood the class well"), the attitude towards STEM topics (4 items such as "I want to program more often"), and how much the students felt that they tried out a lot by themselves (2 items such as "I have tried out a lot by myself").

3.5.3 Individual Competencies

Participants' subjective technical competence (STC) was measured by the STC-questionnaire [16]. It determines person's subjective confidence in his or her own ability to solve technical problems. The short version of the test containing eight items (e.g. "Usually, I successfully cope with technical problems", "I really enjoy cracking technical problems") had to be rated on a six-point scale from 1 (strongly disagree) to 6 (strongly agree). According to Beier's own studies the reliability of the STC short version reaches satisfactory values (Cronbach´s alpha = 0.89). In order to be age sensitive, the wording of some questions was modified.

In addition to the STC we assessed the subjective abilities in arithmetic, geometry and math in general and the subjective competency in dealing with computers. All variables were assessed through bipolar graphical rating scales. Their orientation was randomized on the questionnaires and normalized to the range between 0 and 100 before the statistical evaluation. Low values represent negative and high values represent positive ratings.

4 Results

We analyzed the data using bivariate correlations, χ^2, uni- and multivariate analyses of variance ((M)ANOVA) with a significance level of.05. Pillai values were used for the significance of the omnibus F-tests in the MANOVAs. See Table 1 for an overview of the results.

Four out of 31 questionnaires were not completed by the students. On two questionnaires one answer was missing, on another one two answers were missing and on one questionnaire six answers were missing. To avoid reducing the sample size through the exclusion of questionnaires missing values were substituted by the arithmetic mean of the other participant's answers. This method does not affect the arithmetic mean of the variable in question.

The results section is designed as follows: First, we present the effect of the independent variable gender, second, the effect of the modality of the turtle is shown, and third, the influence of the STC is presented.

4.1 Effect of the Factor Gender

The reported weekly computer usage differed between boys and girls, yielding a significant result ($F(1,29) = 6.19$, $p \leq .05$). On average, the boys use computers 12.8 hours a week, the girls use computers 8.4 hours a week.

The boys had a significantly higher STC (M=73/100 points, SD=13) than the girls (M=55, SD=9, $F(1,29) = 30.67$, $p \leq .01$), corroborating findings of earlier studies

[5, 7, 9]. It is not known however whether this difference results from actual differences in the intrinsic sense of being able to master technological issues or whether girls just answered more modest than the boys on the subjective rating scales provided on the questionnaires. Either ways the results from the girls cannot be compared easily with the results from the boys, as the factor sex is influenced by the STC (η=.717). In the following the STC was used as a covariate to compensate their influence on the factor gender. The median over all students was 64 out of 100 points to be reached at most. It will be used for separating the sample in a group with high and a group with lower STC values (median split).

4.1.1 Self Reported Math and Computer Competencies

The reported arithmetic, geometry and general math competency differed between boys and girls by about 20 points. Girls reported lower arithmetic (M=66, SD=31), geometry (M=57, SD=27) and general math competencies (M=58, SD=18) than the boys (arithmetic (M=85, SD=21), geometry (M=74, SD=28), general math (M=74, SD=30)). The self-reported computer competency differed by 19 points between the boys (M=86, SD=12) and the girls (M=67, SD=27). A MANOVA (taking sex and the turtle's display modality as main factors) showed that these differences are significant (F(4,24) = 3.08, p $\leq$ .05). If the effect of the STC is controlled the differences vanish. We therefore assume that differences are predominantly influenced by differences in the subjective technical competency.

Additionally we found that the nexus between different aspects of math and computer competency differed between boys and girls. We found a linkage between the boys' self reported arithmetic and math competency (r = .539, p$\leq$ .05) and their arithmetic and geometric competency (r = .568, p$\leq$ .05). In contrast, for girls, a significant correlation between the self-reported arithmetic competency and computer competency (r=.571, p $\leq$ .05) was found.

4.1.2 Learning Outcome

To evaluate effects of gender on learning outcomes for each task the students were divided into a group with a lower performance (i.e. three points or below) and a group with good results (four or five points).

Between boys and girls, no difference in the understanding of a given program was found (girls: M=3.2; boys M=3.0). Though, a difference was revealed when girls and boys had to write a software program. With 4.3 points the boys achieved a higher performance than girls (M = 3.6 points). A χ^2 test showed that this difference is marginally significant (χ^2 = 2.761, p= .097 > .05, n.s.).

4.1.3 Class Feedback

The class was equally liked by girls (M=85, SD=13) and boys (M=88, SD=16). Though, the boys found the class more easy to follow (M=88, SD=10) than did girls (M=79, SD=12), revealing a significant effect (F(1,26) = 1.8, p $\leq$ .05).

The attitude towards STEM topics was more positive for boys (M=87, SD=16) than for girls (M=73, SD=20).This difference did not reach statistical significance when the effect of the subjective technical competency was controlled. We assume that attitudes toward STEM topics are carried by gender, but that the main influencing factor is the technical self competence (which is significantly lower in girls compared to boys).

Specifically, the boys had a stronger wish to work on computers more often (M=95, SD=3) than the girls (M=77, SD=24). Also, trial and error behavior and is much more frequent in boys (M=91, SD=13) than in girls (M=75, SD=17), F(2,25) = 4.49, p ≤ .05).

4.2 Effect of the Tangible Artifact

Before the effect of the tangible artifact in regard to class feedback and learning outcome was evaluated, we checked whether both groups are comparable regarding user characteristics. In both groups (the tangible turtle group (8 boys, 6 girls) and the visual turtle group (8 boys, 9 girls)), STC values were equal, reaching on average 64 points. While the reported computer and math competencies did not differ (within 3 points), the reported geometric and arithmetic skills differed by 7 resp. 8 points in favor for the visual turtle group (M=69 vs. M=78). Regarding computer experience, there was a small, but non significant difference in the weekly computer usage (tangible turtle: M=12.4 hours (SD = 5.4); visual turtle: M=9.2, SD=4.9). Overall, we can assume that user characteristics are matched across both experimental groups.

4.2.1 Learning Outcome
The understanding of a given program task was equally large in both groups (M = 3.1). Regarding the task of writing a program the tangible turtle group achieved 4.4 points whereas the visual turtle group achieved 3.7 points. The students were again divided into a group with inadequate results (i.d. three points or below) and a group with good results (four or five points). A χ^2 test showed that this difference is marginally significant (χ^2 = 3.774, p= .052 > .05).

4.2.2 Class Feedback
Both groups were found to have a comparable interest in the class: The tangible turtle group liked the class (M=88, SD=8) about as much as the visual turtle group (M=85, SD=18). The representation of the turtle also had no effect on the perceived simplicity of the class: The visual turtle group (M=83, SD=14) found the class as easy as the tangible turtle group (M=84, SD=9).

The tangible turtle group had a more positive attitude towards STEM topics (M=88, SD=8) than the visual turtle group (M=75, SD=23), though the difference did not reach statistical significance. As this subject area is of special interest for our research we peeked into the individual values of this index. Students of the tangible turtle group were more curious about computer programming (M=88, SD=13) than the visual turtle group (M=71, SD=30), revealing a marginally significant effect (F(1,26) = 3.6, p <.1). Even though students of the tangible turtle group (M=81, SD=14) reported to be inclined to use such a program more often in the future compared to the visual group (M=64, SD=33), the difference missed statistical significance (F(1,26) = 2.7, p = .112 > .05). The same applies for the question if students would like to use more technology in their school education. Even though the tangible turtle group approved this more firmly (M=92, SD=6) than the visual turtle group (M=78, SD=30), no significant effect appeared (F(1,26) = 1.9, p = .179 > .05).

No effect was found for the questions whether the students tried out a lot by themselves. The tangible turtle group answered at about the same level (M=84, SD=15) as the visual turtle group (M=82, SD=19).

4.3 Effect of the Subjective Technical Competence

To understand the effect of the STC on the learning outcome and on class feedback we divided the students along the median ($\tilde{x} = 64$) into a group with low and a group with high STC. Three boys and 13 girls were in the low group and 13 boys and two girls were in the group with high STC values.

We found an almost significant dependency between the STC and the weekly computer usage (($F(1,29) = 4.2$, $p = 0.51 > .05$). The group with high STC used computers 12.6 hours a week whereas the group with low STC had a computer usage of 4 hours less (8.9 hours a week).

4.3.1 Self Reported Math and Computer Competencies

We also found a significant dependency between STC values and the reported computer competency (($F(1,27) = 4.3$, $p \leq .05$). The group of students with high STC rated their computer competency 16 points higher (M=85, SD=11) than the group with low STC (M=69, SD=28).

4.3.2 Learning Outcome

The group with high STC values achieved slightly more points in the task of understanding a given program (M=3.4, $\tilde{x} = 2$) than the group with low STC values (M=2.8, $\tilde{x} = 2$). A χ^2 test showed that this difference is not significant ($\chi^2=2.620$, p=0.106 > 0.05). The STC has indeed an impact on the task of writing a program. The group with high STC values scored over one point better (M=4.5, $\tilde{x} = 5$) than the group with low STC values (M=3.4, $\tilde{x} = 3$). A χ^2 test shows that this difference is significant ($\chi^2 = 4.8$, $p \leq .05$).

4.3.3 Class Feedback

For the reported interest in the class, STC values did not play a significant role. The group with high STC rated their interest with 89 out of 100 points whereas the other group rated it slightly lower with 85 points. Analogically no difference in the perceived difficulty of the class showed up. The High STC group rated the classes difficulty slightly simpler (M=85, SD=12) than the low STC group (M=82, SD=11).

We found no effect of the STC on the question whether the students tried out a lot by themselves. Students with high STC values rated these questions slightly higher (M=86, SD=17) than those with low STC values (M=80, SD=17). An important finding is that the group with high STC had a more positive attitude towards STEM topics (M=84, SD=24) than the low STC group (M=77, SD=13), indicating a significant effect (F(4,24) = 3.6, $p \leq .05$). To assess the effect of the individual items of the four item scale, correlation analyses were used (Spearman's r). We found no connection between the STC and the interest of using more technology in school (r=.11, p>.05). The wish for using computers in school more frequently is connected to the STC (r=.35, p=.055 >.05). A strong correlation existed between STC and the question whether the students would like to computer program more often (r = .450, $p \leq .05$) and the question whether the students had become curious about computer programming in general (p = .45, $p \leq .05$).

Table 1. Overview of the factors gender, tangible artifact and subjective technical competence

	Factor Gender		Factor Turtle		Factor STC	
	Boys	Girls	Visual	Tangible	Low	High
Weekly Computer Usage [h]	12.8	8.4	9.2	12.4	8.9	12.6
Subjective technical competence	73	55	64	64	-	-
Reading a Program (0..5)	3.0	3.2	3.1	3.1	2.8	3.4
Writing a Program (0..5)	4.3	3.6	3.7	4.4	3.4	4.5
Number of students	16	15	17	14	16	15

5 Discussion and Conclusions

5.1 Using Robots in Computer Science Education

The empirical evidence suggests that tangible turtles have mild advantages over turtles presented visually. We saw no differences regarding the understanding of a given computer program, however the learning outcome was increased for writing a computer program. The data suggests that using tangible robots encourages a positive attitude towards STEM topics, although no considerable effects were found.

These findings are similar to those from the Roberta project that states that robots put computer science into a meaningful perspective. But we must also acknowledge that using robots in computer science education won't solve the problem that too few school students strive for a career in computer science and that the participation of women in STEM and computer science in particular is highly asymmetrical.

5.2 Gender and Technology

It's not surprising that we found a huge gender gap with regard to the subjective competencies in math, computers and general subjective technology competence. We could confirm that these subjective feelings affected the learning outcome, especially for the task of writing a computer program. These results fit into Busch's study that stated that women have lower self-efficacy that negatively impacts performance outcomes. This was especially true for the complex task of writing a program. Our results contradict the promising findings that gender differences in school math are dissolving in the USA were boys and girls archive similar results in standardized math tests [17]. The girls in our study reported much lower competencies that are accompanied by lower performance in writing computer programs.

A cautionary note refers to the fact that our findings rely on subjective measures that might be affected by systematic under- or overestimations. Although this effect was statistically controlled, gender differences were still present. It is an interesting finding that girls and boys showed different profiles: Boys found that math related aspects belong together, whereas girls saw computers and math strongly related.

We were surprised that such gender differences were present in the seventh grade and that these differences influenced the girls' performance negatively. Future research must focus on the question of when girls and boys drift apart and what methods can hinder that. We firmly believe that the STC is a mediating factor for interest and success in scientific education and career building.

5.3 Subjective Technical Competence as a Mediator for Successful Learning

We learned that the STC is an important mediator for interest in STEM topics and successful learning. It also strongly influences the self attributed competencies in geometry and dealing with computers. We found out that high STC leads to better scores in understanding a given computer program and especially writing a computer program. We also revealed that the STC is usually much lower for girls than for boys. Therefore measures to increase interest in STEM topic and measures to level gender differences in science and technology must focus on building subjective competences for boys and girls in addition to just teaching hard facts. The belief to be able to master technology seems to have a great impact on the actual performance. This study also reveals that measures to increase interest in science and technology should start earlier than in the seventh grade, because the STC in this age is already much lower for girls than for boys.

5.4 Limitations

Our study was carried out with a limited number of participants so a balancing of girls and boys with similar STC was not feasible. Additionally most of the variables assessed are based on subjective statements so the analysis is prone to systematic over- or underestimations of the students.

To generalize the results from this study we need to complement the attitudes towards science and math classes with performance outcomes. Here, the link between attitudes and school success (i.e. grades) are of interest as well as cognitive abilities and aptitudes which are connected to science and math performance, as e.g. problem solving abilities, speed of information processing, or spatial abilities [9, 18].

Class feedback was very positive in both conditions. We suspect however that this was mainly caused by the alternation from school routine, absence of the pressure to perform and the small group size. To hedge the results the study needs to be repeated under realistic conditions, i.e. with regular class sizes, over a longer period of time and with grading throughout the course.

5.5 Outlook

During this study we developed a ten item computer programming self-efficacy scale modeled after a previous scale [19]. We found a meaningful connection between our scale and the standardized STC, as students with a high STC also had a high computer programming self-efficacy. Its internal consistency was satisfying for the first iteration of the scale (Cronbach's $\alpha = .615$). However, it is not ready yet for practical use as its clarity is blurred by the combination of multiple items into one scalar variable. As self-efficacy is an important factor in career theories further research on creating a usable scale for assessing school students programming self-efficacy has to be developed.

Our study revealed that teaching scientific understanding must start way before the seventh grade. Too few research projects focus on the scientific and technical education of younger children to help constructing scientific understanding and the feeling that the adults of tomorrow are able to master technological devices with ease.

References

[1] Leszczensky, M., et al.: Bildung und Qualifikation als Grundlage der technologischen Leistungsfähigkeit Deutschlands. Bericht des Konsortiums Bildungsindikatoren und technologische Leistungsfähigkeit. Studien zum deutschen Innovationssystem, Nr. 8-2008 (2008)

[2] Schinzel, B.: Informatik und Geschlechtergerechtigkeit in Deutschland - Annäherungen. In: Gender and Science. Perpektiven in den Natur- und Ingenieurwissenschaften. Bielefeld, pp. 127–145 (2007)

[3] Bandura, A.: Self-efficacy: Toward a unifying theory of behavioral change. Psychological Review 84(2), 191–215 (1977)

[4] Liu, L., Grandon, E.E.: How performance and self-efficacy influence the ease of use of object-orientation. In: Proceedings of the 36th Hawaii International Conference on System Science (HICSS 2003), Big Island, HI, January 2003, pp. 327–336 (2003)

[5] Brosnan, M.J.: The impact of computer anxiety and self-efficacy upon performance. Journal of Computer Assisted Learning 14, 223–234 (1998)

[6] Arning, K., Ziefle, M.: Understanding age differences in PDA acceptance and performance. Computers in Human Behavior 23(6), 2904–2927 (2007)

[7] Busch, T.: Gender differences in self-efficacy and attitudes toward computers. Journal of Educational Computing Research 12, 147–158 (1995)

[8] Zeldin, A., Pajares, F.: Against the Odds: Self-Efficacy Beliefs of Women in Mathematical, Scientific, and Technological Careers. American Educational Research Journal 37(1), 215–246 (2000)

[9] Ziefle, M., Jakobs, E.-M.: Wege zur Technikfaszination. In: Sozialisationsverläufe und Interventionszeitpunkte. Springer, München (2009)

[10] Petersen, U., et al.: Roberta Abschlussbericht, BMBF Projekt (2007), http://www.iais.fraunhofer.de/fileadmin/images/pics/Abteilungen/AR/PDF/Abschlussbericht_Roberta_2007-11-21.pdf (accessed March 2009)

[11] Westram, H.: Schule und das neue Medium Internet - nicht ohne Lehrerinnen und Schülerinnen. Thesis (Dr.), Universität Dortmund (1999)

[12] Schelhowe, H.: Interaktivität der Technologie als Herausforderung an Bildung. Zur Gender-Frage in der Informationsgesellschaft. Ruhr Universität Bochum (2004), http://www.ruhr-uni-bochum.de/fiab/pdf/jahrbuch/j17a5.pdf (accessed March 2009)

[13] Hartmann, S., Schecker, H.: Bietet Robotik Mädchen einen Zugang zur Informatik, Technik und Naturwissenschaft? – Evaluationsergebnisse zu dem Projekt Roberta. Zeitschrift für Didaktik der Naturwissenschaften, Jg. 11 (2005)

[14] Monroy-Hernández, A., Resnick, M.: FEATURE: Empowering kids to create and share programmable media. Interactions 15(2), 50–53 (2008)

[15] Papert, S.: Mindstorms:Children, Computers and Powerful Ideas. Basic Books, Inc., New York (1980)

[16] Beier, G.: Locus of control when interacting with technology (Kontrollüberzeugungen im Umgang mit Technik). Report Psychologie 24, 684–693 (1999)

[17] Hyde, J., et al.: DIVERSITY: Gender Similarities Characterize Math Performance. Science 321(5888), 494–495 (2008)

[18] Ziefle, M., Bay, S.: How to overcome disorientation in mobile phone menus: A comparison of two different types of navigation aids. Human Computer Interaction 21(4), 393–432 (2006)

[19] Cassidy, S., Eachus, P.: Developing the computer user self-efficacy (CUSE) scale: investigating the relationship between computer self-efficacy, gender and experience with computers. Journal of Educational Computing Research 26(2) (2002)

The Difficulty of Programming Contests Increases

Michal Forišek[*]

Comenius University, Bratislava, Slovakia
forisek@dcs.fmph.uniba.sk

Abstract. In this paper we give a detailed quantitative and qualitative analysis of the difficulty of programming contests in past years. We analyze task topics in past competition tasks, and also analyze an entire problem set in terms of required algorithm efficiency. We provide both subjective and objective data on how contestants are getting better over the years and how the tasks are getting harder. We use an exact, formal method based on Item Response Theory to analyze past contest results.

1 Introduction

It is a well-known consensus in the community around programming contests that the difficulty of these contests progressively increases. For example, Verhoeff et al. mention this observation in [23] as a part of the motivation to have a Syllabus for the International Olympiad in Informatics (IOI).

In this article, we try to give exact, objective arguments that this is indeed happening. We analyze the "internals" of competition tasks (such as covered topics and required algorithm efficiency). Also, we analyze the raw competition data (such as detailed results, submission logs, etc.) that can be used to show whether tasks are getting harder.

We will now give an overview of prior research in this area. The age at which children start practicing for programming contests is decreasing. Kelevedjiev and Dzhenkova [10] mention that in many countries the age at which children start with programming and programming contests is already as low as ages 11 to 12. Thanks to this early start the contestants are able to cover more areas of computer science during their preparations. In accord, the set of topics used in contest tasks is growing. For example, Manev [14] notes that tasks on graphs became common in all levels of contests after such tasks were used at the IOI. Each year there are proposals of new task types that push the boundary of the scope of programming contests still further – for recent examples see [3,21,16].

But even if we restrict ourselves to a fixed set of topics, there will still be both easier and harder tasks around, and the harder ones are more and more common. Kiryukhin in [12] describes the development of the Russian Olympiad in Informatics. One important point mentioned by Kiryukhin is that only after

---

[*] This work was supported by the grant VEGA 1/0726/09.

J. Hromkovič, R. Královič, and J. Vahrenhold (Eds.): ISSEP 2010, LNCS 5941, pp. 72–85, 2010.

the introduction of modern, efficient computers around the year 1995 the organizers were able to use test inputs large enough to evaluate algorithm efficiency
with sufficient precision. Hence the best contestants became motivated to learn,
design and implement better, more efficient algorithms.

Note that from Kiryukhin's observation it follows that only for the last approximately 15 years algorithm efficiency plays a significant role in programming
contests. In other words, since 1995 the focus in programming contests started to
shift from "implement a correct algorithm" towards the much harder "implement
a correct algorithm that is as efficient as possible".

In [18, Section 7] Revilla et al. discuss the issue of task difficulty. They were not
able to find a satisfying way to determine the difficulty of tasks in the University
of Valladolid (UVa) Online Judge [17]. They state that difficulty is subjective,
and while most people agree on the ends of the spectrum, the difficulty levels
of intermediate tasks are almost impossible to establish. For the book [19] the
difficulty of the selected tasks was estimated manually.

For past IOI tasks there are several publications related to the scope of this
article: Kiryukhin and Okulov [13] (in Russian) manually analyze and classify
the past tasks, Verhoeff [22] provides a different clarification and also ranks
the difficulty of tasks based on the percentage of contestants who managed to
"fully solve" the task (i.e., score at least 90% of possible points). As claimed
by Verhoeff, this metric only describes how hard the task was for the set of
contestants who were solving it – but it is not sufficient to compare the difficulty
of tasks from different years.

Kemkes et al. [11] introduce Item Response Theory (IRT) as a tool that can
be used to evaluate the difficulty of competition tasks, and they use it to analyze
scoring of tasks from past IOIs. Their methods were further developed by the
author of this article in [5,6].

1.1 Goals of the Article

In this article we present detailed evidence (both quantitative and qualitative)
for the following claims:

1. The set of topics covered by task statements and solutions is growing.
2. The topics previously considered difficult now start to appear in contests
   designed to be "easier" (such as categories for younger students).
3. **The difficulty of tasks in programming contests is increasing.**
4. The skill level of both top and average contestants is increasing.

We will mostly focus on Claim 3. However, it is important to realize that claims
3 and 4 are closely related. In Section 2 we illustrate this on an example.

1.2 Overview of the Article

In Section 2 we show that raw scores, more precisely IOI medal boundaries, do
not carry sufficient information to argue about task difficulty.

In Section 3 we address the first two goals of our article, and additionally we show that in tasks based on the same computational problem there is a trend towards requiring more efficient algorithms.

In Section 4 we present the results of our international survey that shows a strong correlation between the year in which a task was set and its perceived difficulty.

In Section 5 we address the fourth goal of this article, showing data that the skill level of both the very best and the average contestants is gradually increasing. The main claim of this article logically follows from the data presented in the previous three sections.

Finally, in Section 6 we give a short note on how Item Response Theory can be used to obtain data that will give us even more information on task difficulty.

2   IOI Medal Boundaries

A naïve attempt to prove that the difficulty of programming contests increases would simply lead us to examine the scores – with the hypothesis that we should see a steady decrease.

In Figure 1 we show the medal boundaries at the IOI in the years where automated grading was used.[1] For clarification: the three lines denote the score achieved by the last contestant that was awarded a gold, a silver, and a bronze medal respectively. (This corresponds to the top 8.3%, top 25%, and top 50%.)

Clearly, there is no visible decrease in the scores, more precisely, no significant negative correlation between the year and the medal boundaries. Quite on the contrary, the scores tend to be pretty balanced throughout the years, with seemingly easier and harder years alternating. As we already mentioned (and as we show in later Sections), the actual reason is that two trends occur at the same time – not only are the tasks getting harder and harder, but also the contestants are getting better and better.

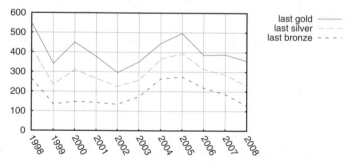

Fig. 1. IOI medal boundaries in years 1998 to 2008

---

[1] In all years except for 1998 and 2000 the theoretical maximum was 600 points. In 1998 the maximum was 700, the scores are scaled. In 2000 each contestant was given 100 free points, hence those 100 points were subtracted from the scores.

The conclusion we can draw from this simple example is that if we want to argue about task difficulty, we can not do it without addressing the contestants' skills at the same time. Or, from a statistical point of view, we will not be able to draw any conclusions from contest results only, unless we assign contestant names to the scores or add some other additional data source.

3 Task Topics and Algorithm Complexity

We now show several related trends: In Section 3.1 we show that difficult concepts such as dynamic programming become more and more common in solutions, and such concepts now occur even in tasks supposed to be "easy". In Section 3.2 we show that tasks built upon the same computational problem now usually require more efficient algorithms than in the past. Finally, in Section 3.3 we show that each year there are new tasks requiring new, more complex algorithms.

3.1 Task Topics in TopCoder Contests

TopCoder Inc. organizes programming contests since 2001. Since 2003, there are regular single-round competitions, called Single Round Matches (SRMs) that all use the same format: The contestants are separated into two (approximately equally large) divisions according to their current *rating*.[2] The stronger division is called Division 1 (Div1), the weaker one is Division 2 (Div2). In each Division, the contestants are given three tasks to solve. The tasks are labeled "easy", "medium" and "hard" according to their expected difficulty. Tasks in Div1 are different (and harder) than tasks in Div2.

In the task archive each task has a set of labels assigned by its author. The labels list the general methods used to solve the task, such as "dynamic programming", "geometry", or "graph theory". These labels and also detailed results of all past matches can be obtained from the official data feeds [20].

We focused on four of the more interesting labels. For each of these, we plotted a graph that shows, for each difficulty level in each division, the percentage of tasks that had this label in each half-year since 2003. These plots are shown in Figure 2. The main points that can be observed by reading these plots:
– The number of "easy" tasks based on brute force and simulation decreases.
– The number of tasks that require difficult concepts increases, and such tasks are becoming more and more frequent in easier difficulty levels.

In Subfigure 2a we see the steady decline of hard tasks that can be solved by brute force. This is most clearly visible in Div1 hard tasks (the hardest task in each round), where the percentage of tasks solvable by brute force decreased steadily since 2006, and actually reached zero in 2009.

A similar trend is visible in Subfigure 2b. In 2005 simulation tasks were still pretty common even as Div1 hard tasks, but since 2006 the number of such tasks never exceeded 10%, and it was zero in 2009. Almost the same is true for

---

[2] The rating is a TopCoder-calculated numeric estimate of the contestant's skill. Ratings currently lie in the range $[0, 4000)$, higher is better.

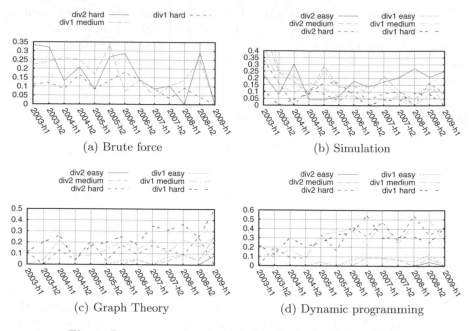

Fig. 2. Percentages of topic-related tasks in TopCoder's SRMs

Div1 medium tasks. On the other hand, more and more of the easiest tasks are simulation-based.

Subfigure 2c shows the percentage of graph theoretical tasks. Even in the second half of 2004 almost no Div1 hard tasks contained graph theory. Ever since, this number is steadily growing, and in 2009 it reaches 50%. Also observe that several years ago graph theoretical tasks were used as the more difficult problems only. Until 2005 such tasks were only used as the harder two tasks in Div1, and as the hardest task in Div2. In 2006 the first such tasks appeared as Div1 easy and Div2 medium tasks, and the percentage of such tasks is growing steadily ever since. In 2009 the first task on graphs was used as Div2 easy.

Finally, in Subfigure 2d we can observe similar trends for tasks that require dynamic programming. The popularity of such tasks was growing between 2003 and 2006, and is more or less constant since. Again, we can observe that since 2005 dynamic programming is finding its way into the easier task levels. In 2008 a dynamic programming task was used as Div2 easy for the first time.

3.2 Efficient Algorithms in UVa Contests

For technical reasons, the number of elements in data structures used in Top-Coder tasks is limited to 50. E.g., if the input is an adjacency matrix of a graph, the graph can only have at most 50 vertices. This does, to some extent, make it impossible for problem setters to enforce the use of the most efficient algorithms for a given task. E.g., the most efficient algorithms for single-source shortest paths in a graph have time complexity roughly proportional to $N \log N$, but for

$N = 50$ even an $O(N^3)$ algorithm will do fine. However, other contests lack this limitation, and in such contests we can also observe that as years progress, it is more and more common to use test data sizes that require the contestants to implement more efficient algorithms.

We analyzed of tasks in the "Contest" section of the UVa Online Judge [17] archive. We focused on several well-known computational problems, for each of them we found tasks based on it, and examined their input sizes. When analyzing the tasks, two helpful resources were the manual classifications of UVa Online Judge tasks done independently by Greve [7] and by Naverniouk [15].

The results are shown in Table 1. In all four cases, the evolution of maximum instance sizes clearly shows the progress towards requiring asymptotically more efficient algorithms – with the most recent versions requiring (almost) optimal ones known.

Table 1. Maximum instance sizes in UVa contest tasks

(a) Single-source shortest paths

id	name	date	N	M	comment
10269	Adventure of Super Mario	2002-04-20	1 000		requires preprocessing
10342	Always Late	2002-07-27	200		2nd shortest walk
10603	Fill	2004-01-10	80 000	240 000	solvable via BFS as well
10740	Not the Best	2004-10-16	1 000	10 000	k shortest walks
10917	Walk Through the Forest	2005-09-24	1 000		number of shortest paths
10986	Sending email	2006-01-21	20 000	50 000	
11367	Full Tank?	2007-12-01	100 000	1 000 000	state: city and fuel amount
11635	Hotel booking	2009-07-18	10 000	100 000	additional complications

(b) Convex hull

id	name	date	N	comment
10065	Useless Tile Packers	2001-01-19	100	
10173	Smallest Bounding Rectangle	2001-09-01	1 000	$O(N^2)$ postprocessing needed
10652	Board Wrapping	2004-05-22	2 400	
11072	Points	2006-08-12	100 000	followed by point-in-polygon queries
11096	Nails	2006-09-21	100	
11168	Airport	2007-02-17	10 000	non-trivial postprocessing
11626	Convex Hull	2009-06-13	100 000	

(c) Longest increasing subsequence

id	name	date	N	comment
10131	Is Bigger Smarter?	2001-06-29	1 000	
10534	Wavio Sequence	2003-07-26	10 000	two computations needed
10635	Prince and Princess	2004-03-13	62 500	non-trivial preprocessing needed

(d) Strongly connected components

id	name	date	N	M	comment
10319	Manhattan	2002-06-29	120	400	solving 2-SAT
10731	Test	2004-09-25	26		
11324	The Largest Clique	2007-10-27	1 000	50 000	
11504	Dominos	2008-09-27	100 000	100 000	

In all tables, N is the main instance size: the number of vertices/sequence elements/points. Wherever the input is a graph, M is the limit on its number of edges. If omitted, $M = O(N^2)$.

3.3   New Task Types by Year

In this Section we attempt to show how the set of topics used in contest tasks is growing in time. In order to do so, we created a list of topics that, according to our opinion, form a representative sample of different algorithms and areas of Computer Science. For each of the topics we attempted to find the earliest contest and task that involved this topic. We then arranged the topics into a list ordered by the year of the first appearance.

Obviously, it is not humanly possible to take all international contests into account. The list we present is based on the following contests:

- International Olympiad in Informatics [8] 1989–2009
- The set of ACM ICPC contests at UVa Online Judge [17] 2000–2009
- Internet Problem Solving Contest [9] 1999–2009
- TopCoder competitions [20] 2000–2009

One important set of contests that is missing in the list is the set of World Finals of the ACM ICPC [1] – so far we did not have sufficient resources to analyse these tasks. (Also, this analysis is only possible to some extent, due to the fact that the official test data used for the contest is destroyed after the contest.)

1991: grammars and rewriting systems (IOI: S-terms)
1995: theoretical task; process communication (IOI: Printing),
interactive task (IOI: Wires and Switches)
1996: a two-player interactive game (IOI: A game),
optimal job scheduling (IOI: Job processing)
1997: optimization tasks and approximation algorithms (entire IOI)
1999: open-data task format (entire IPSC 1999),
code analysis and white-box testing (IPSC: Coins)
2000: game played in multiple submissions (IPSC: Trolls),
querying an unknown sequence (IOI: Median)
2001: Fenwick trees (IOI: Mobiles),
meet-in-the-middle search (IOI: Double crypt),
max. weight bipartite matching (BUET/UVA Oriental C.1: Bob Laptop),
a-star heuristic search (2001 Regionals Warmup Contest: 15-Puzzle problem)
2002: reducing a 2d task to independent 1d tasks (IOI: Utopia),
dynamic programming exponential in some dimension (TC: PinballLanes),
general matching, Chinese postman problem (U. of Waterloo June Contest)
2003: programming non-traditional computation models (IPSC: begin 4 7 add)
2004: reducing runtime from $O(N!)$ to $O(2^N)$ via DP (TopCoder: KiloManX)
2006: pen and paper cryptoanalysis (IPSC: Encoded messages),
modern cryptoanalysis (IPSC: h4x0r t3h c0d3),
page faults in caching (IPSC: Librarian)
2007: minimum cost maximum flow (TC: RadarGuns)
2008: low-level data representation (IPSC: Comparison mysteries)
2009: regular expressions processing (IPSC: Muzidabutur)

4   Subjective Task Difficulty Rating

For the purpose of obtaining useful data on how contestants perceive task difficulty, we carefully selected a set of 12 tasks using the following methodology: All the tasks come from different years of the ACM ICPC [1] Northwestern European regional contest (NWERC). More precisely, for each year between 1997 and 2008 inclusive we examined the results of the contest and picked the median task[3] according to the number of teams that solved it.[4] The rationale behind this choice was that the median task should reasonably well represent the difficulty of the problem set.

The statements of those 12 tasks (without any indication of their origin) were presented as a survey to active contestants. The goal in the survey was to estimate the relative difficulty of those 12 tasks. Details on the survey formulation are given in Appendix A.

We got 33 answers to the survey. The respondents that filled in the survey were from over 20 different countries on 5 different continents, and all of them are successful participants in international programming contests.

The results of the survey are presented in Table 2. Tasks are labeled 0 to 11 according to the year in which they were used (0 is oldest). The left half of Table 2 contains the raw answers: the number in row X and column Y is the number of respondents who think task X is harder than task Y. Shaded cells are those for which the opposite opinion received strictly less votes.

The right half contains the same data, but we only count the 26 non-anonymous respondents for which we know their TopCoder ratings (see Section 3.1 for explanation of ratings). We used the ratings as weights of their votes. In this way we gave a higher significance to votes by better contestants. The totals are in thousands, rounded to the closest one.

Table 2. Survey of NWERC task difficulties. Left part: raw, right part: weighted.

	0	1	2	3	4	5	6	7	8	9	10	11	0	1	2	3	4	5	6	7	8	9	10	11
internet/0		0	0	1	2	0	0	1	0	2	0	0		0	0	1	3	0	0	1	0	3	0	0
space/1	27		20	11	18	8	8	5	3	12	5	5	46		40	16	32	12	12	8	2	18	9	9
papergirl/2	23	5		8	11	6	6	5	3	8	4	6	41	5		10	17	7	8	6	4	8	3	9
railroads/3	21	14	16		19	12	12	9	6	8	6	8	40	31	37		35	28	27	15	10	17	10	16
dates/4	24	6	15	6		8	7	6	4	9	5	5	41	12	28	10		14	14	10	6	11	8	9
floors/5	24	15	15	12	17		10	7	7	13	6	8	42	28	32	17	30		14	10	10	17	9	14
boss/6	23	14	15	10	13	12		4	5	9	6	8	42	27	30	16	23	25		5	9	14	8	16
taxicab/7	22	17	15	15	18	14	15		15	13	14	13	36	34	33	28	32	29	30		24	21	26	24
tantrix/8	24	20	21	18	22	15	16	8		19	13	11	40	39	41	32	36	30	31	14		31	22	22
setstack/9	21	10	13	13	18	10	11	7	5		10	9	39	21	29	25	33	23	21	11	7		20	18
escape/10	26	20	17	16	20	15	15	7	11	13		14	43	38	35	29	35	31	30	12	18	19		25
mobile/11	22	15	16	13	18	12	12	7	10	12	7		39	27	32	23	32	24	22	12	16	20	14	

---

[3] Ties were broken by the total number of submissions (or equivalently, number of incorrect submissions, less means easier), and if the problem set contained $2N$ tasks then the N-th easiest task was selected.

[4] Even more precisely, for the years 1997 to 1999 only summary ranklists were available. For these contests we selected the median task by hand.

There clearly is a significant correlation between the year in which the task was used and its perceived difficulty. (A huge majority of shaded cells is below the diagonal, which means that the tasks from later years were labeled as more difficult.) The only significant difference between the two tables is in the row/column 9 corresponding to the "SetStack" task from NWERC 2006. This task is deceiving: it seems solvable by a plain simulation, but this is false. The replies in our survey match the actual results of the contest, where 31 out of 42 teams attempted to solve this task, but only 9 of them solved it. In the right half of Table 2 we see that the higher rated responders found this task harder.

Finally, we note that the trend of increasing median task difficulty seems to stall in last five years, hence it would be interesting to repeat the survey after several years to find out more about this new trend.

5   Comparison of Performance in Different Years

We already argued that when discussing task difficulty we have to take into account the skill levels of the contestants. However, the converse is not necessarily true: we may be able to compare the skill levels of contestants in different years. Where does the symmetry break? For the contestant, a few years can mean all the difference in the world – a few years of practice (or inactivity) can strongly influence the contestant's skill in solving the tasks. However, a task after several years is precisely the same task. Hence the tool that can help us compare skill levels of the contestants in different years is a task that was given in different years to (ideally) two disjoint sets of contestants. We were able to discover multiple such situations and we analyzed them.

5.1   TopCoder Tasks

In general, TopCoder does a pretty good job in avoiding repeating the same task. Still, we were able to find several pairs of almost identical tasks in past TopCoder matches. Statistic data on these tasks is given in Table 3. The column "opened" is the number of contestants who attempted to solve the task, "solved" is the number (and percentage) of those who managed to submit a correct solution. The next two columns give the fastest and the average solving time, respectively; the average is only computed over all contestants who solved the task. The last column is the number of contestants who competed in both rounds.

Notable facts: By observing the number of solvers and the average solving times for Div1 tasks, the rate of their improvement is obvious – both for the top and the average Div1 contestants. On the other hand, the performances on the easy Div2 tasks are almost identical in early 2007 and late 2008. Essentially the same task was used as a hard task in 2007 and as a medium in 2009.

5.2   Slovak Selection Camp

Tasks are sometimes reused in the Slovak selection camp – a week-long camp for approximately the best 10 contestants in the Slovak Olympiad in Informatics,

Table 3. Pairwise similar tasks used in TopCoder contests over the years

task	date	level	opened	solved		fastest	average	overlap
Layoff	Mar 2003	div1 hard	138	11	(7.97%)	0:21:45	0:35:51	10
Terrorists[a]	Jan 2007	div1 hard	385	102	(26.49%)	0:02:44	0:18:13	
InstantRunoff	Dec 2003	div1 easy	160	85	(53.13%)	0:07:07	0:23:01	10[b]
InstantRunoffVoting	Mar 2008	div1 easy	583	469	(80.45%)	0:02:48	0:16:59	
InstantRunoff	Dec 2003	div2 med.	187	37	(19.79%)	0:16:08	0:44:42	0[b]
InstantRunoffVoting	Mar 2008	div2 med.	759	220	(28.99%)	0:07:52	0:34:03	
Graduation	Jun 2004	div1 hard	120	2	(1.66%)	0:20:13	0:30:07	12[c]
SharksDinner	Jul 2007	div1 hard	379	51	(13.46%)	0:10:27	0:24:18	
CountPalindromes	Feb 2007	div1 hard	379	11	(2.90%)	0:25:47	0:42:47	63[d]
PalindromePhrases	Apr 2009	div1 med.	447	61	(13.65%)	0:07:21	0:41:19	
Palindromize	Jan 2007	div2 easy	569	371	(65.20%)	0:01:42	0:20:54	39
ThePalindrome	Dec 2008	div2 easy	751	454	(60.45%)	0:02:47	0:19:59	

[a] Layoff is a plain maximum flow task, Terrorists requires finding multiple minimum cuts.

[b] Two more people competed in Division 2 in 2003, and in Division 1 in 2008.

[c] None of the two solvers in 2004 took part in 2007.

[d] Only 24 of these managed to solve the task. The other 37 solvers did not participate in 2007.

where the IOI team is selected. The data for those tasks is presented in Table 4. For each task, we normalized the score to 100 points. The column "top 4" contains the sum of the normalized scores of the four best solvers for that task, the column "OK" is the number of participants that scored at least 90% of points for the given task. In the column "medals" you can find the medals acquired by the Slovak IOI team that year – G, S, B standing for gold, silver, and bronze, respectively.

Out of the 11 tasks, only in 5 of them was the best result obtained in the most recent year the task was used. From this observation, we have to conclude that in the data from the selection camp we do not observe the increase of skill

Table 4. Tasks reused in the Slovak selection camp

task	year	OK	top 4	medals
wine	2009	7	400.0	SSB
	2007	7	400.0	SSB
syr	2003	4	400.0	GSB
	2001	4	400.0	GGSS
sally	2009	8	400.0	SSB
	2003	1	347.6	GSB
dazdovky	2007	1	222.0	SSB
	2004	1	253.3	SSBB
	2001	2	269.2	GGSS
jazdci	2007	1	188.6	SSB
	2003	0	105.7	GSB
	2001	2	337.5	GGSS
klinec	2007	0	292.0	SSB
	2003	3	362.7	GSB
guards	2007	0	166.7	SSB
	2003	0	130.7	GSB
junior	2006	0	147.7	SSS
	2001	1	360.0	GGSS
domceky	2006	2	308.0	SSS
	2002	4	390.0	GSSB
prufer	2005	6	390.0	GGGG
	2001	4	384.0	GGSS
gule	2004	1	215.0	SSBB
	2002	0	119.3	GSSB

levels that was clearly visible on the international scale. We see two factors that could contribute to this contradiction significantly.

First, the Slovak OI is so small that it is not statistically significant. With only about 100 students competing in this contest each year, the skill levels of the top ones can vary wildly from year to year. And judging by the Slovak results at the IOI, they indeed do vary. As for the data from the selection camp, kindly note that for 9 of the 11 tasks[5] the best result in the selection camp occurred in the year in which Slovakia obtained the best medals.

Second, in Section 5.1 we were not able to consider one important factor – the ages of the contestants. Note that the TopCoder data set contained contestants of all ages, whereas the selection camp is always attended by secondary school students only. At the moment, we do not have sufficient data to look into this, and we consider this to be an interesting question for further research.

6   Evaluating Task Difficulty Using IRT

Item Response Theory [2] is a modern testing theory that gives us a set of new tools to analyze contest results. Among other things, this theory allows us to model tasks with different difficulties. More precisely, the two-parameter logistic model is a suitable model to describe programming contest tasks. In this model, each contestant c is assumed to have a scalar ability level θ_c, and each task t is assumed to have two scalar parameters a_t and b_t that describe its difficulty. In this model, the probability that a contestant with ability θ_c will solve a task with parameters a_t, b_t is $Pr(\theta, a, b) = 1/\left(1 + e^{-a(\theta - b)}\right)$. (Note that b is equal to the value θ for which $Pr(\theta, a, b) = 0.5$, and a is the derivative in this point.)

In [5] the author of this paper developed a framework how Item Response Theory can be applied to the results of programming contests. Using this framework, it should be possible to make a thorough analysis of the past contests, compute maximum likelihood estimates of their task parameters, and use this data to draw conclusions about the relative difficulty of those contests.

Due to space restrictions we will not attempt such analysis here. Instead, we will just limit ourselves to a single example of such results. Using the method described in [5] we computed the maximum likelihood estimates for contestant and task parameters describing all tasks from TopCoder's SRMs in 2007 and 2008, and all contestants that took part in those matches.

In Figure 3 we show the plots of item characteristic curves for all Div1 hard tasks used in matches in 2007 (left) and 2008 (right). The x axis on both plots represents ability levels (the entire range being $[-5, 5]$), the y axis is probability of solving a given task, and each task corresponds to one of the plotted logistic functions. In layman's terms, the further to the right a curve is, the harder the

---

[5] The task "sally" is one of the other two. For this task the conditions in 2003 and 2009 were not identical. This is a task solvable by brute force. In 2003 the task was used with a strict time limit, and implementing good pruning was necessary to get a full score. In 2009 all correct solutions got a full score.

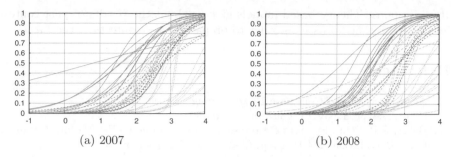

(a) 2007   (b) 2008

Fig. 3. Item characteristic curves for hard tasks in TopCoder matches

task – because this means that the contestant's ability has to be large in order
to have a chance to solve the task.

For easier readability of the plot, we ordered the tasks according to their b_t
parameter, grouped them into four quarters, and plotted each quarter using a
different color and pattern. Note that the parameter b_t is equal to the ability
level necessary to have a 50% chance of solving the task. In other words, b_t is
the x coordinate of the point where the logistic curve crosses the line $y = 0.5$.

Observe the line $y = 0.5$ to see the values b_t in Figure 3. The first quartiles of
b_t are almost the same (2.23 in 2007, 2.3 in 2008). However, note that in 2008
almost all b_t in the first quarter exceed 2. The median task in 2007 is already
significantly easier than in 2008 (2.6 vs. 2.8), and the trend is even more visible
at the third quartile (2.82 vs. 3.23).

7   Conclusion

Programming contests are still growing in popularity. The community of contestants is growing, and this puts an added pressure on their preparation (and
performances). The contestants are getting better and better each year. As a consequence, the tasks in the contests they solve must become harder and harder,
to be able to distinguish between the top contestants.

In this article we give sufficient evidence that this process is still happening.
When will it stop? In our opinion, we should see parallels between the area of
programming contests, and all of Computer Science – which is still one of the
most rapidly developing sciences. As long as the boundaries of Computer Science
grow at the current pace, the contests will mimic this growth, and incorporate
the new discoveries into new, even more challenging tasks – thereby preparing a
future generation of scientists who can push the boundaries even further.

We see many open questions and directions for further research in this area.
To list some: Is this process a good thing, or should we attempt to stop it? Aren't
the contests already too difficult? How does the gradually increasing difficulty
impact newcomers? To what extent does the increase in difficulty also require an
increase in mathematical skills of the contestants? What is the relation between
the age of contestants and their skill levels? How much more insight can the

Item Response Theory based models give us about the results of past contests, and how can they be used in order to prepare better contests in the future?

References

1. ACM International Collegiate Programming Contest, http://cm2prod.baylor.edu/ (accessed 2008)
2. Baker, F.B., Kim, S.: Item Response Theory: Parameter Estimation Techniques. CRC, Boca Raton (2004), http://edres.org/irt/baker/
3. Burton, B.: Breaking the routine: events to complement informatics olympiad training. Olympiads in Informatics 2, 5–15 (2008)
4. Fenwick, P.: A New Data Structure for Cumulative Frequency Tables. Software – Practice And Experience 24, 327–336 (1994)
5. Forišek, M.: Theoretical and Practical Aspects of Programming Contests. PhD thesis, Comenius University (2009)
6. Forišek, M.: Using Item Response Theory to Rate (Not Only) Programmers. Olympiads in Informatics 3, 3–16 (2009)
7. Greve, M.: UVA toolkit (2009), http://uvatoolkit.com/
8. International Olympiad in Informatics, http://ioinformatics.org (accessed 2009)
9. Internet Problem Solving Contest, http://ipsc.ksp.sk/ (accessed 2009)
10. Kelevedjiev, E., Dzhenkova, Z.: Tasks and training the youngest beginners for informatics competitions. Olympiads in Informatics 2, 75–89 (2008)
11. Kemkes, G., Vasiga, T., Cormack, G.: Objective Scoring for Computing Competition Tasks. In: Mittermeir, R.T. (ed.) ISSEP 2006. LNCS, vol. 4226, pp. 230–241. Springer, Heidelberg (2006)
12. Kiryukhin, V.: The Modern Contents of the Russian National Olympiads in Informatics. Olympiads in Informatics 1, 90–104 (2007)
13. Kiryukhin, V., Okulov, S.: Methods of Problem Solving in Informatics: International Olympiads. Izdatelstvo BINOM (2007) (in Russian)
14. Manev, K.: Tasks on graphs. Olympiads in Informatics 2, 90–104 (2008)
15. Naverniouk, I.: Igor's UVa Tools (2009), http://shygypsy.com/acm/
16. Opmanis, M.: Team Competition in Mathematics and Informatics "Ugāle" – finding new task types. Olympiads in Informatics 3, 80–100 (2009)
17. Revilla, M., et al.: University of Valladolid (UVa) Online Judge (2009), http://uva.onlinejudge.org/
18. Revilla, M., Manzoor, S., Liu, R.: Competitive Learning in Informatics: The UVa Online Judge Experience. Olympiads in Informatics 2, 131–148 (2008)
19. Skiena, S., Revilla, M.: Programming Challenges. Springer, Heidelberg (2003)
20. TopCoder, Inc.: Algorithm Data Feeds (2009), http://www.topcoder.com/wiki/display/tc/Algorithm+Data+Feeds
21. Truu, A., Ivanov, H.: On Using Testing-Related Tasks in the IOI. Olympiads in Informatics 2, 171–180 (2008)
22. Verhoeff, T.: 20 Years of IOI Competition Tasks. Olympiads in Informatics 3, 149–166 (2009)
23. Verhoeff, T., Horváth, G., Diks, K., Cormack, G.: A Proposal for an IOI Syllabus. Teaching Mathematics and Computer Science 4, 193–216 (2006)
24. Verhoeff, T., Horváth, G., Diks, K., Cormack, G., Forišek, M.: IOI Syllabus for IOI 2009 (2009), http://www.ioi2009.org/GetResource?id=32

A   Survey on NWERC Tasks

The survey contained a list of 12 tasks, ordered randomly. The order was different for different people. The purpose of this randomization was to ensure that the respondents have no preconceptions on the difficulty of the tasks.

These are the exact instructions given to the people taking the survey:

Your task is to **order** the following ACM ICPC problems based on **how difficult** you find them.

- Click the problem name to read the problem statement. *Please read the problem statements and think about the solutions!*
- Consider the overall difficulty of solving the task, from opening the problem statement to submitting a correct solution.
- Use the dropdown boxes to assign numbers 1 to 12 to the problems.
- You *may assign the same number* to problems that have approximately the *same difficulty* according to you.
- You also may *only rate a subset of the tasks* and omit those where you are not certain. Even such feedback is valuable.
- The absolute values of numbers you use to rate the tasks do not matter. The only thing that matters is: *If you give task A a **smaller** number than task B, you think that A is **easier** than B.*
- The tasks are shown in random order. This order changes when you reload the page. Hence it is a good idea not to reload the page while you are entering your choices.

Didactic Games for Teaching Information Theory*

Michal Forišek[1] and Monika Steinová[2]

[1] Comenius University, Bratislava, Slovakia
forisek@dcs.fmph.uniba.sk
[2] ETH Zurich, Switzerland
monika.steinova@inf.ethz.ch

Abstract. We developed a set of didactic games and activities that can be used to illustrate and teach various concepts from Information Theory. For each of the games and activities we list the topics it covers, give its rules and related information, describe our practical experiences and give an overview of its scientific background. We also discuss the proper ways to integrate these games into the knowledge acquisition process.

1 Introduction

It is a well-established fact that games do belong in the classroom. Various aspects in which games contribute to the educational process were researched and supporting evidence for this fact has been given in many publications.

In recent years, the main focus in this area was on computer games. For the general setting, the review [4] notes that people acquire new knowledge and complex skills from game play, and that the computer games can teach higher-order thinking skills such as strategic thinking, interpretative analysis, problem solving, plan formulation and execution, and adaptation to rapid change. McFarlane et al. [16] point out that games provide a forum in which learning arises as a result of tasks stimulated by the content of the games, knowledge is developed through the content of the game, and skills are developed as a result of playing the game. Recently, Interactive Software Federation of Europe sponsored a major study [24] on the use of computer games in schools in Europe. Other relevant publications on general use of computer games in classrooms include [18,23].

It would only seem natural that out of all school subjects, computer science would be the one where including computer games would provide the most benefits. As an example, Van Emde Boas [3] mentions how strategic computer games can be used to teach concepts related to information systems modelling and relational databases. Essentially, the player is forced to extract useful information from the game and build an abstract model in order to succeed.

Computer games are becoming well established in the teaching process. However, we are strongly convinced that pen and paper, or even physical, games and

---

* This work was partially supported by FILEP grant 351 of ETH Zurich.

J. Hromkovič, R. Královič, and J. Vahrenhold (Eds.): ISSEP 2010, LNCS 5941, pp. 86–99, 2010.

activities can have a stronger impact in the classroom – especially if they are tailored to the topic we intend to teach.

The use of such games in teaching mathematics is well researched. See Hatch [6] for an overview, and Rowe [17] for a concrete example. For use of classroom games in teaching concepts from statistics and economy, see multiple publications by Holt and various coauthors, especially [11,12].

In teaching computer science, the flagship is the book by Bell et al. [2], where the authors give twenty games and activities that can be used to illustrate and teach various concepts from computer science. Ginat [5] focuses on (combinatorial) mathematical games and shows that the development of strategies for such games exercises both rigor and heuristic reasoning – which both present key competences in real-life problem solving. Hill et al. [10] use puzzles and games to teach various concepts in operating systems, and Levitin and Papalaskari [14] show that the reasoning students use to solve pen and paper puzzles can be used to introduce and illustrate algorithm design techniques.

2 Role of Games in Knowledge Acquisition

Whenever we decide to use a didactic game or activity in the classroom, we should see the game as a tool, and we should be aware of the goals this tool helps us to accomplish. We find it very helpful to think about the knowledge acquisition process in terms of the Theory of Generic Models (TGM), as developed by Hejný [8,9]. (See also Sfard [22] for a different but similar theory.) According to Hejný's theory, the process of knowledge acquisition should follow the following six steps:

1. motivation for knowledge acquisition
2. gathering experiences, forming separate models
3. generalization
4. discovery of the generic model
5. abstraction, crystallization of the knowledge
6. abstract, automated knowledge

To briefly illustrate these steps, consider a small child that learns how to count. Separate models include the pieces of knowledge *two candies plus one candy are three candies* and *two cars plus one car are three cars*. In the generalization process, the child then forms a generic model *anything can be counted on fingers, and two fingers plus one finger are three fingers.*

This generic model covers many separate models the child previously developed. But it is still bound to concrete objects, which may hinder the child in further development of this knowledge. In such case, this knowledge needs to be abstracted – in our example towards the symbolic equation $2 + 1 = 3$. And only then the usage of this knowledge can become automated – so that the child can apply the equation $2 + 1 = 3$ without needing to think about any concrete objects.

And where exactly do classroom games enter the picture?

Classroom games and activities of the type we present in our paper may play several important roles. The two most significant ones correspond to the first two steps of the knowledge acquisition process described above.

As we already mentioned in the overview of past research, games can be an excellent way to involve and activate the students. When playing the game, the students encounter problems they need to solve, and their desire to play the game well is a strong motivational factor to do so.

But, in our opinion, even more important is the contribution of the games towards the second step: forming separate models. For example, in Section 5 we present a game where one of the players controls a "robot" using just a few simple commands. On their own, the players will develop a separate model of information coding and using a transmission channel in the context of the game.

We conclude this section with a brief note on the third step: generalization, progress towards a generic model. For each of the games we provide a subsection with background information. This subsection usually contains the generic models the students will later form in their studies of Information Theory. It is possible for the teacher to reveal some or all of these after the students finish playing the game – but this is not required, and we recommend not to do it.

The most exciting part of the knowledge acquisition process is the arrival at the fourth step – the discovery of a generic model. As Hejný showed, for the best didactic effect it is important for the students to make this discovery on their own. By revealing the generic model too early, the teacher is taking the extasy of this discovery away from the students. Hence we usually (and with much success) apply the following approach:

- Play a suitable game.
- At some later point in time, introduce a new concept.
- Thanks to playing the game, the students now have two separate models (one from the game, one from the teacher's introduction), and hence it is easier for them to start the generalization process on their own.

3   Overview of the Paper

In this paper we focus on teaching concepts from Information Theory, as started by Shannon [20] in 1950. For a recent publication that gives an overview of all topics we discuss in this paper, see for example Cover [1].

The main contribution of this paper is a set of didactic games and activities that can be used to introduce, illustrate, or even teach various concepts in Information Theory. All of the games and activities we present were "field-tested" under various conditions – mostly on camps for talented secondary school students in Slovakia. For each game we include a section where we describe our practical experience with the game, and another section where we discuss the scientific background behind the particular game.

The games and activities in this paper illustrate the following main concepts:

- **The definition of information.** Information we get is equal to our surprise. The less probable the observation, the more information we get.
- **Information can be measured.** Asking a yes/no question may give us a bit of information – but we have to ask "smart" questions.

- **Information is relative.** Same event may give different amounts of information to different people.
- **Transmission channel.** Used to send information "from A to B".
- **Data coding.** When transferring data, we often need to pre-process it into a form that can be sent via the transmission channel.
- **Data compression.** Reasons, methods, lower bound = information content.
- **Redundancy of languages.** In the natural language each letter carries less than 2 bits of information. Why is this a good thing?
- **Error correction.** Hamming distance: To guarantee that we can correct transmission errors, no two possible transmissions may be too similar.
- **Covert channels.** Misusing environment to transfer information.

In the following Section we give a short note with our variation on the well-known "Twenty questions" game. Each of the remaining Sections describes one of our games and activities.

4   Guess the Sentence

Concepts: The definition of information. Information can be measured. Information is relative.

Bell et al. [2] nicely explain how the common "Twenty questions" game[1] can be modified to introduce the basics of Information Theory. However, we must point out two issues that we feel are not handled correctly in their version.

First, when guessing an entire sentence, Bell et al. give the instruction: "the sentence should be guessed one letter at a time, from left to right". Clearly, the motivation for this has been Shannon's original experiment [19]. However, that choice is not appropriate in our context, where our primary goal is to measure information content. By introducing such arbitrary restrictions we are influencing the outcome of our measurements.

We can nicely show this on an example from our practical experience: We had players guessing sentences in the "letter by letter" way. This game was often frustrating: a word was clear after guessing its first few letters, and the player felt that she is wasting questions by being forced to confirm each and every letter in order. These questions gave the player no new information, hence the number of questions did not correspond to the information amount.

The version in which arbitrary questions are allowed is more precise in measuring information content; and also more entertaining from the player's point of view, as it offers a wide spectrum of strategies – semantic, syntactic, or combined in various ways.

The second point missing in the presentation by Bell et al. is the relativity of information. When playing this game, we often have several players guess the same sentence independently of each other, and compare their scores. In such situations we must keep in mind that information is relative – and make the secret sentence "equally improbable" for each of the players.

---

[1] The activity is called "Twenty guesses" in their book.

Alternately, we may use this game in the opposite way: to illustrate the point that information is relative. For example, we had players guess the sentence "Cuban rum is made from sugar cane".[2] On average, players who did not know this fact needed about twice as many questions as those who did.

5   Soy-Sugar-Glue

Concepts: Data coding. Transmission channel. Covert channels.

5.1   Rules

The game is played by two players, called *the robot* and *the navigator*. During the actual game the robot has eyes covered by a blindfold, and he is not allowed to speak. Apart from that, the robot is allowed to do anything he wishes to – but he does not know the goal he is supposed to achieve. On the other hand, the navigator is told the goal when the game starts, but she is only allowed to say three different words. We customarily use the words "soy", "sugar", and "glue" (hence the name of the game), but any three distinct words will do.

Multiple robot-navigator pairs can be given the same goal, and then compete which of them can achieve it first.

5.2   Materials and Preparation

After the rules are explained to the players, and robot-navigator pairs are formed, the players need to be given time to prepare – they need to devise a way how the navigator can manipulate the robot using just those three words. The exact preparation time varies with age, for the age group 15 to 18 usually 30 minutes were sufficient.

It is helpful to give the players some simple example goals, so that they have something to practice on during the preparation time. A list containing some goals we used (including their approximate difficulty) is given in Appendix A.

5.3   Strategies and Practical Experience

There is no clear optimal strategy to this game. There are plenty of ways in which the commands can be mapped to various actions. Usually, the best strategies are those that are flexible enough (i.e., give the navigator a sufficient freedom), but at the same time are systematic enough for the robot to actually remember them. When devising a strategy, it is useful to visualize the robot as a string puppet – we need to be able to make his various body parts move in various directions.

One common example of a good strategy is to use commands starting with "soy" to describe body parts, and commands starting with "sugar" to describe movements. For most goals given in Appendix A, it is sufficient to be able to describe the basic body parts (entire person, head, body, left/right arm/leg), and the six basic directions (up/down, left/right, forward/back).

---

[2] In Slovak: Kubánsky rum sa robí z cukrovej trstiny.

We often find that the players' designs mimic existing interfaces that they encountered in practice. It is quite common to see an "undo" command that reverses the last action. One particularly lovely interface developed by a pair of players was a special command that put the robot into a "select body part" mode. The robot was using his right index finger to point at various body parts, and the navigator was able to select and confirm the right one. In this way, just a few commands were sufficient to be able to select virtually any body part.

On some occasions we had a pair of players attempting to "cheat" – instead of mapping commands to the robot's actions, they mapped commands to the letters of the alphabet (e. g., using Morse code or binary). Then, instead of navigating the robot, the navigator simply sent him the goal using their chosen encoding. However, this type of "cheating" is easily prevented. Note that the robot is blindfolded, and therefore unaware of its surroundings. Any task that requires orientation in the surroundings can not be accomplished in this way.

A different (and in some settings acceptable) way of "cheating" is to use intonation, volume, speed of speech and various other effects in order to communicate more than just the meaning of the command – dismay in the navigator's voice can signalize that the robot is doing something wrong, speed of speech can be reflected in the speed (and amount) of the robot's movements, etc.

5.4   Background and Insights

The rules of this game were developed by the organizers of camps for talented youth, probably around the year 1990. This entertaining game nicely illustrates the concept of a transmission channel, and the need to encode information. The information is the actual content we want to transmit, whereas the transmission channel is something we have available. Information coding is a necessity – we have to process the information we have into a form that can be sent along the transmission channel.

Additionally, there are other, more subtle aspects related to Information Theory in this game. A prominent one is the concept of covert channels (see Lampson, [13]). This is precisely what is happening in the last type of "cheating" we mentioned – in addition to the official transmission channel, there is a possibility to transmit more information using a different, hidden channel that originally was not supposed to be used in this way.

6   Reconstructing Damaged Text

Concepts: Redundancy of languages. Error correction.

6.1   Rules

– *Non-interactive version 1* (word completion):
  The teacher picks a suitable paragraph of text, damages it using one of the methods outlined below, and then highlights about ten words. Each of the students is given a copy of the resulting text. The students are then given a

fixed amount of time (we use 10 minutes) to fill in as many of the highlighted words as they can. Only exact matches count. (Deciphering other words is not required, but it may help to understand the text.)

- *Non-interactive version 2* (question answering):
  Instead of highlighting words, the teacher prepares a set of questions about the original text. The students try to answer as many of them as possible.
- *Interactive version*:
  If this activity is performed in a computer lab, we can use its interactive version. Again, the student is given a text to decipher and understand, only this time the text is displayed in an application that allows the student to control the amount of damage to the text. Initially, this amount is set to maximum. The student's goal is to decrease the amount of damage until he can read most of the text (and possibly prove it in the same way as in one of the non-interactive versions).

6.2   Materials and Preparation

An online script. We implemented a simple script that can be used to play the interactive version, and also to prepare materials for the interactive version. Currently it supports English and Slovak. Location:
`http://ksp.sk/~misof/damage/`

Different ways to damage text. First of all, we note that in all the activities in this section we only consider effects that damage letters – spaces and punctuation remain untouched.[3] When damaging a text we have to choose two different things: how to select the letters to damage, and how to damage them.

The basic choices for selection of letters are *periodic* (every n-th letter is damaged) and *random* (each letter is damaged with the same probability). Another significant choice is to damage *vowels only*.

The letters we pick up to be damaged can either be *hidden* (replaced by a new "blank" symbol, e. g. '_'.), or *corrupted* (replaced by a different letter). There are multiple ways how to pick the new letter to replace the old one; we considered two of them: *uniform* (each letter equally likely) and *natural frequencies* (each letter as likely as it normally is in the given language).

Preparing suitable materials. For this activity we usually picked a paragraph of text (about 1000 characters long). Care should be taken to avoid "unfair" texts that contain a significant amount of information only known to some of the students. For the non-interactive version 1, we recommend to pick the 10 words by hand, and only to do so after the text is damaged. For the non-interactive version 2, when phrasing the questions it is important to keep two goals in mind: First, the questions should be as unambiguous as possible, ideally with single-word correct answers. Second, the questions we ask must reveal some information about the damaged text; care must be taken to minimize this amount of information.

---

[3] Note that as a consequence the word lengths are always preserved.

6.3 Strategies and Practical Experience

An entertaining way to do this activity is to pick one fixed type of text damage, and try to discover what is the largest amount of damage you can have if you want to be able to reconstruct (most of) the original text. The interactive version is intentionally designed with this purpose in mind.

The non-interactive versions can be seen as an "offline approximation" of the interactive version. A good way to achieve a similar effect as in the interactive version is to play the game with 2 to 5 different texts, all damaged using the same method, but with different amounts of damage.

We had a test group of 23 solvers play the non-interactive versions of the game with 5 different Slovak texts. The outcomes are summarized below.

In the non-interactive version 1 (word completion) the most successful solvers actually tried to understand as much of the text as possible. However, there was a significantly large group of successful contestants that only focused on filling in the highlighted words (and their surroundings, if necessary). This behavior may have been influenced by their perception of the time constraint.

In the non-interactive version 2 (question answering) only those people who managed to understand a substantial portion of the text had a chance to actually answer some of the questions. However, despite our best efforts we often received comments from the solvers that some particular questions helped them slightly in trying to understand the text.

If only vowels were damaged, most of our solvers were able to reconstruct almost the entire original text, even if all vowels were hidden or corrupted. If we allow arbitrary letters to be hidden, the amount of damage we can afford becomes decreases. In our test group, a major part of a text with 40% letters hidden was successfully restored by most solvers, but another text with 55% letters missing already proved too challenging for many (but not all!) of the solvers. For the non-interactive version 2 we used a text in which 35% of letters were replaced by new letters (with the same probability distribution as in the Slovak language). This proved to be by far the most difficult text to restore. Summary data on all the performances is given in Appendix B.

6.4 Background and Insights

Redundancy of natural languages evolved as people needed to understand each other even in the presence of noise. It is precisely the redundancy that allows our brain to correct minor mistakes, and fill in the missing parts in the speech we hear. This feature of languages is not only well understood, but also exploited. For example, in inherently noisy environments such as aviation [7], intelligibility of spoken communication is increased using standardized phraseology – such as *negative* instead of *no*.

Error correction codes in computer science work in very much the same way: redundancy is what makes them possible. However, redundancy is just a necessary condition, not a sufficient one. This can also be shown in the context of our game: Consider the sequence: `Peter has a _at.` What exactly does he have?

A cat? A hat? Something else? We can not be sure. And in this way we can discover the sufficient condition for an error correcting code: In order to be able to correct errors in the transmission, every two possible messages must have a sufficiently large edit distance.

For many languages most of the information in written text is carried by consonants – and in some languages vowels are commonly omitted in the written form. This can also be shown using our game: vowels account for approximately 40% of letters in English text, but an English text with all vowels removed is easier to reconstruct than an English text with random 40% of letters removed.

7   Text Message Game

Concepts: Data compression. Redundancy of languages. Error correction.

7.1   Rules

This game is played by several competing pairs of students. Before playing the game, the teacher has to choose a suitable paragraph of text.

- *Version 1* (blackening letters)
  From each pair, one student is given a copy of the teacher's text, typeset in a monospace font (i.e., all letters have the same width). The student now may blacken as many letters as she wants to. For each such letter her team gains a point. Letters must be blackened completely, so that they are not readable. Once the first student is finished, she hands over the edited text to her partner. He has to reconstruct the original text. For each incorrect word, his team loses 20 points. The team with most points in the end wins.
- *Version 2* (text message shorthand)
  As above, one student is given a copy of the teacher's text. She has to rewrite it onto a clean sheet of paper, using only uppercase letters and spaces.[4] Her goal is to use as few symbols as possible. Once she is done, the provisional score of her team is calculated as the number of characters she saved. Again, as above, her partner then has to reconstruct the original, losing 20 points for each incorrect word.

7.2   Material

As in Section 6, the teacher has to select a "fair" paragraph of a text, so that extra knowledge of the topic of text does not help the students in better reconstruction. A good text should only consist of letters, spaces and basic punctuation. Punctuation may be removed by the teacher if desired. Note that in version 1 the monospace font is necessary in order to prevent guessing missing letters from their width.

---

[4] Optionally, numbers and basic punctuation may be allowed.

7.3   Background and Insights

When writing text messages on cell phones, it is natural to "compress" the text in order to fit in more information. In many languages including English, this leads to significant changes in the written form of the language. One significant type of changes is using phonetic equivalents, e. g., "u", "2", and "w8" instead of "you", "to/too", and "wait". Another type of changes is simply leaving out unnecessary letters, e. g., "msg" instead of "message". This game is built upon this process, which should come natural to most students.

The reason why we decided to present two versions is that they offer a trade-off, and it is up to the teacher to pick the more suitable one. The advantage of the first version is that it is purely mechanical, less time consuming, and incredibly easy to evaluate. On the other hand, the second version is more attractive. It is closer to the "original" text messaging, it gives the players an opportunity to be creative and look for ways to compress the text even more. But to do this they will require more time. Also, we have to check whether the sheets produced by the first players are correct.

8   Knocking Game

Concepts: Data coding. Transmission channel. Data compression.

8.1   Rules

The game is played by a teacher and a pair of students.

The students are put into two adjacent rooms with a closed door inbetween. The teacher picks a simple short sentence (4-5 words) in the students' native language, and tells this sentence to the first student. The first student has some time to prepare. Then, at a moment that is announced to the second student, a five minute period is started. During this period, the student that knows the sentence has to transmit it to his friend in the other room. However, the only allowed way of communication is that the first student is allowed to knock at the door.

The obvious first goal is to transmit the sentence without any mistakes. The second, more challenging goal is to do this using as few knocks as possible.

8.2   Materials and Preparation

No materials are needed and no preparation (other than finding a suitable location to play the game) is required on the teacher's part.

Between explaining rules and actually transmitting the message the students need to be given time to devise the best strategy they can find. Regardless of the setting we estimate that a minimum of 30 minutes is necessary, but we recommend even more. One good scenario is to explain the rules at the end of a lesson and then play the game during the next lesson (which takes place in a day or so).

8.3   Strategies and Practical Experience

The obvious basic strategy is to transmit the sentence character by character.

The first attempt is usually using unary coding: 1 knock is A, 2 is B, etc. Students that start with this strategy usually quickly realize that it is quite straining (and error prone) to make more than 20 knocks just to transmit a single letter that is close to the end of the alphabet.

Sometimes, especially with younger students, we then encountered crude optimizations, such as shifting more frequent letters (e. g., vowels) to the beginning of the alphabet – so that they are represented by less knocks.

The usual next conceptual step the students discover is to use various knocking patterns to decrease the number of knocks. An especially popular case is to use two knocking patterns, clearly inspired by modern computer mice: "knock" and "double-knock".

By far, the most common variation of this approach is that the students assign the values 1 and 0 to knock and double-knock, and then for each letter in the sentence they transmit its index in the alphabet. A slightly more efficient (but less common) variation is to use Morse code, with knocks and double-knocks representing dots and dashes.

The optimal variation of this approach considers the possible knock/double-knock patterns (or even includes triple- and quadruple-knocks), orders them according to their total number of knocks and assigns them letters in this order.

Additionally, we sometimes encountered crude compression methods. The most popular method was to use text message style abbreviations (see Section 7.3 for details).

Our experience is that only the best secondary school students are able to proceed beyond this point. The following strategies will more commonly be devised by university students.

When actively trying to minimize the number of knocks, one can discover that the double-knocks actually waste knocks. In the same way short and long pulses were used in telegraphs, we can use pauses of various length *between* the knocks. Mathematically speaking, this observation lowers the number of knocks necessary to transmit n binary symbols from an expected $3n/2$ to $n + 1$.

The best solution we saw students discover so far was performed by two students who were both successful participants in the International Olympiad in Informatics. The approach they finally developed was based on dictionary coding – they created a list of common Slovak words, assigned them numbers based on their frequency, and then only transmitted those numbers, using the technique from the previous paragraph. Their solution needed less than 50 knocks to transmit an approximately 30-letter sentence.

8.4   Background and Insights

The door in this game (the one between the students, used to transmit the message by knocking on it) is actually a model of an *transmission channel*. As you cannot knock letters, there is the need to *encode information* into a form that can be sent by this channel. Furthermore, the rules of the game reward those who can devise an efficient way how to encode this information and *compress* it into as few knocks as possible.

(There is one significant difference between the game and the real life settings: in the game we are *not* trying to minimize transmission time, the compression is measured in a different way.)

The knock/double-knock approach shows how an analog channel can be used to transport a digital signal. The transmissions that use knocks and double-knocks are essentially using a binary alphabet to transmit their message, and timing is used to separate knocks belonging to different symbols.

The idea with pauses between knocks can actually be pursued further – we can introduce pauses of different lengths. When regarded from the theoretical perspective, this corresponds to picking a larger alphabet for the transmission channel,[5] However, this approach has an upper bound on k: we must be able to transfer the message within the time limit, and this may be impossible for large k and limited precision of measurements achievable by the receiving student.

See Appendix C for a table with a comparison of the above strategies.

9   Conclusion

We presented a set of various games and activities that involve concepts from Information Theory. Most of these activities are original, developed by the authors of this paper. All of these activities are simple enough to be explained to (and played by) 12 years old children – but on the other hand, they have a deep background, and can be challenging even for college students. For each activity, we give our insights into its background, and where possible we provide helpful observations from our practical experience. It is our hope that these activities will, in the future, help the next generations understand Information Theory as something nice, simple and natural.

References

1. Cover, T.M., Thomas, J.A.: Elements of information theory, 2nd edn. Wiley-Interscience, New York (2006)
2. Bell, T., Fellows, M., Witten, I.: Computer Science Unplugged... Off-line activities and games for all ages (1998), http://csunplugged.com/
3. Van Emde Boas, P.: Games in the Classroom. In: OOPSLA 1999 workshop (1999)
4. Federation of American Scientists: Harnessing the power of video games for learning. Summit on Educational Games (2006)
5. Ginat, D.: Elaborating heuristic reasoning and rigor with mathematical games. SIGCSE Bull. 39, 32–36 (2007)
6. Hatch, G.: A rationale for the use of games in the mathematics classroom. Topic Issue 19, NFER (Spring 1998)

---

[5] This decreases the number of symbols to transmit by a constant factor. With a k-symbol alphabet we need approximately $\log_k 2^n = n/\log_2 k$ symbols to transmit n bits of information.

7. Hawkins, F.H.: Human Factors in Flight, 2nd edn. (2001); reprinted by Ashgate
8. Hejný, M.: Knowledge without understanding. In: Proceedings of the international symposium on research and development in mathematics education, pp. 63–74 (1988)
9. Hejný, M.: Understanding and Structure. In: CERME3 – Conference on European Research in Mathematics Education 3, Group 11 (2003)
10. Hill, J., Ray, C., Blair, J., Carver, C.: Puzzles and games: Addressing different learning styles in teaching operating systems concepts. In: SIGCSE 2003, pp. 182–186 (2003)
11. Holt, C.A., Anderson, L.: Classroom Games: Understanding Bayes' Rule. Journal of Economic Perspectives 10(2), 179–187 (1996)
12. Holt, C.A., Capra, M.: Classroom Experiments: A Prisoner's Dilemma. Journal of Economic Education 31(3), 229–236 (2000)
13. Lampson, B.W.: A Note on the Confinement Problem. CACM 16, 613–615 (1973)
14. Levitin, A., Papalaskari, M.-A.: Using puzzles in teaching algorithms. SIGCSE Bull. 34, 292–296 (2002)
15. Mahoney, M.: Refining the Estimated Entropy of English by Shannon Game Simulation (1999), http://cs.fit.edu/~mmahoney/dissertation/entropy1.html
16. McFarlane, A., Sparrowhawk, A., Heald, Y.: Report on the educational use of games. Shelford Studio, 46 Whittlesford Road, Little Shelford, Cambridge (2002), http://www.teem.org.uk/publications/teem_gamesined_full.pdf
17. Rowe, J.: An experiment in the use of games in the teaching of mental arithmetic. Philosophy of Mathematics Education Journal 14, 1–16 (2001)
18. Sandford, R., Williamson, B.: Games and Learning, A Handbook from Futurelab. Futurelab, Harbourside, Bristol (2005), http://www.futurelab.org.uk/resources/publications_reports_articles/handbooks/Handbook133
19. Shannon, C.E.: Prediction and Entropy of Printed English. Bell Sys. Tech. J. 30, 50–64 (1950)
20. Shannon, C.E.: A Mathematical Theory of Communication. Bell Sys. Tech. J. 27, 379–423, 623–656 (1948)
21. Stur Language Institute (Jazykovedný ústav L'. Štúra), Slovak Academy of Sciences: Slovak national language corpus – prim-3.0-public-all (2007)
22. Sfard, A.: On the dual nature of mathematical conceptions: reflections on processes and objects as different sides of the same coin. Educational Studies in Mathematics 22, 1–36 (1991)
23. Virvou, M., Katsionis, G., Manos, K.: Combining Software Games with Education: Evaluation of its Educational Effectiveness. Educational Technology & Society 8, 54–65 (2005), http://www.ifets.info/journals/8_2/5.pdf
24. Wastiau, P., Kearney, C., Van den Berghe, W.: How are digital games used in schools? Complete results of the study. European Schoolnet, EUN Partnership AISBL (2009), http://games.eun.org/upload/gis-full_report_en.pdf

A   Soy-Sugar-Glue: Possible Goals

Here we list some possible goals for this game, grouped by difficulty. Easy goals are suitable as examples for practice.

Easy: Pick up an object. Rotate on the spot. Get down on all four.
Medium: Touch your nose. Clap hands. Score a goal with a football.
Hard: Perform a forward roll. Find a bottle, open it and drink from it.
Very hard: Jump forward. Give a non-playing person a kiss on the cheek.

B   Reconstructing Damaged Text: Examples, Statistics

In Figure 1 we show two short paragraphs of damaged text. In the first paragraph
two words are highlighted. For the second paragraph, answer the following question: "What feeling does the player experience?" (Note that the word "player"
is clearly readable, hence we are not giving away too much information.)

l_ssr_om g_m_s _nd $\boxed{\text{_ct_v_t__s}}$ [1] _f th_ typ_ w_ pr_sent _n o_r p_p_r
m__ pl_y $\boxed{\text{s_v_r_l}}$ [2] _mportant r_l_s.

```
Tois game cas often frrsetatbng: r ward wws tlaar after ruetsing
ias firnt feu neutirs, and tho player felt trat she is waseini
queseires wryn tutssirg the rsst of she wyrd.
```

Fig. 1. Damaged texts: 75% vowels missing ; 25% letters replaced

In Table 1 we present the numbers of successful solvers (out of 23) in our field
test of the non-interactive versions of this activity. For version 1, the columns
correspond to the 10 words we highlighted, for version 2 these are answers to
our 10 questions.

Table 1. Aggregate results of the "Damaged text" activity

version	damage type	q1	q2	q3	q4	q5	q6	q7	q8	q9	q10
v. 1	100% vowels replaced	22	23	22	23	22	22	22	22	22	13
v. 1	100% vowels hidden	23	19	21	20	20	22	22	19	23	20
v. 1	40% letters hidden	21	22	20	22	22	22	22	17	10	10
v. 1	55% letters hidden	20	21	20	20	18	13	8	21	20	16
v. 2	35% letters corrupted	22	7	21	15	22	4	0	20	9	12

C   Knocking Game: Comparison of Strategies

In Table 2 we present a comparison of different strategies for the English sentence
"The rabbit hole went straight on like a tunnel." (38 letters).

Table 2. Comparison of strategies for the Knocking game

method	knocks
unary coding	442
unary coding based on English letter frequencies	262
plain binary coding of letter indices	206
Morse code	123
optimal binary coding based on letter frequencies	119
plain binary; gap lengths instead of double-knocks	144
dictionary coding; binary using gap lengths	85
dictionary coding; base-4 using gap lengths	50

Collaborative Initiatives for Promoting Computer Science in Secondary Schools

Irene Glendinning[1] and Margaret Low[2]

[1] Faculty of Engineering and Computing, Coventry University
ireneg@coventry.ac.uk
[2] Warwick Manufacturing Group, University of Warwick
m.j.low@warwick.ac.uk

Abstract. This paper presents a model for local collaborative working to support school teachers and their pupils in computer science. In the city of Coventry in the UK a network of representatives from various local government organizations, universities, professional bodies and schools has been working together for mutual advantage to organize and support initiatives that encourage and facilitate the teaching of computer science and information and communications technology in schools. Major events to date include a day conference, workshops and seminars for ICT teachers and a series of web design competitions for local schools. The paper discusses advantages and conflicts of a networking approach for improving understanding between different agencies working towards a common goal and concludes with plans for further collaborative ventures.

Keywords: ICT in schools, Local initiatives for ICT, collaborative working ICT in schools, Professional body input to ICT in schools, personal networking approach to ICT in schools, Vertical interaction between tertiary and secondary education in ICT, encouraging entry to computer science at HE level, Informatics competitions in schools.

1 Introduction

In Coventry an effective network exists for sharing ideas and information related to ICT and computer science education. The common purpose uniting members is the desire to provide a better classroom experience for pupils in local schools studying computing and information and communication technologies (ICT) in the hope that this will encourage them towards a career in computer science. The network provides an effective means of communication between IT and education professionals.

A three year study, reported by Kozma & McGhee in 2003, focused on the relationship between ICT use and innovative practices in classrooms across 28 countries, one aspect of the study examined how the use of ICT changed the organization of the classroom [1]. They recognized that ICT does break down barriers between students, subject areas, students and teachers, but that most of the change stayed within the classroom. Only a few cases were found where teachers and students used ICT to collaborate with external actors, for example scientists, professors, and business people.

In contrast with the study by Kozma and McGhee, a key strength of the network described here is the vertical interaction between tertiary education and secondary education, informing and supporting the network participants. In addition this initiative

J. Hromkovič, R. Královič, and J. Vahrenhold (Eds.): ISSEP 2010, LNCS 5941, pp. 100–111, 2010.

concerns sharing good practice across the educational and organizational boundaries of the different partners.

For simplicity of expression the term ICT is used in this paper to encompass the broader spectrum of informatics, computer science, computer communications and information technology, as relevant in the context of secondary school curricula.

2   The Network

2.1 Background

Serendipity was behind the formation of the network rather than deliberate intent and design. The authors met though professional membership of the British Computer Society and act as representatives in the local BCS Coventry branch for the two universities in Coventry: University of Warwick and Coventry University.

The main impetus for action began in 2004 when members of a local Coventry branch of BCS decided to run a competition for local schools. A sub-committee of branch members convened to plan, organize and implement their first competition. Advice and guidance was provided by the Oxfordshire branch of BCS who regularly run their own schools' competition. Their support was crucial to the success of the first BCS Coventry branch competition in 2006. The ethos for mutual support and networking with a range of agencies was established from this time. Later both the Coventry and Oxfordshire branches were asked to provide advice to the Jersey BCS branch for their first schools competition, which successfully operated in 2006-7.

A larger Coventry BCS sub-committee (the schools committee) was established in 2007 to oversee the more substantial second Coventry branch schools competition. At the June 2008 competition final there was strong demand from participating schools for an annual competition. The schools committee took the decision that it was not feasible to operate a competition annually, mainly because of the demands on volunteers and resources. However it was agreed that it would be sensible to maintain and develop the effective contacts established during the interim year, not least the network of ICT teachers. In response some support events for teachers were organized during 2008-9, as described later in this paper.

During the same period several new links were established through peripheral and work-related activities of the schools committee members. From a series of regular meetings common objectives and symbiotic relationships between various agencies emerged. The new members of the network acted as advisers and contributors to decisions and events during the year, serving to enrich and inform the activities and operational aspects. Early in 2009 the schools committee formally convened to begin to organize the third Coventry Branch schools' competition, again informed with guidance and advice from wider network members.

2.2 Network Partners

Without wishing to diminish the contribution of other people, the key to the success of all events to date has been the interest and enthusiasm from a strong team of volunteers who are all members at the local Coventry branch of BCS. The BCS nationally recognizes the value of links between industry and education - at all levels- primary, secondary and tertiary - and supports research in this area. The BCS Education forum highlighted problems facing teachers in their debate 'Feeding the pipeline: What can we do with schools' as reported in the article by Preston and Mellors – Bourne 2008 [2].

Coventry is fortunate to have two thriving universities which are both strongly represented and actively encouraging the work of the network. Coventry University's Faculty of Engineering and Computing, University of Warwick's department of Computer Science and Warwick Manufacturing Group have provided facilities, resources and high level expertise to host and organize events and training.

Former and current members of Coventry and Warwickshire education departments (representing Coventry city and surrounding county and region) that have specific responsibility for providing ICT support and advice to schools are important members of the network. These people have knowledge of current policies and plans for the future relating to schools curriculum, pedagogy, resources and funding within their particular domain. They can draw upon their own network of contacts when a specific resource is needed or question is raised.

In the UK there is a government initiative to promote Science, Technology Engineering and Mathematics (STEM) within schools. The STEM coordinator for Coventry and Warwickshire is a pivotal member of the local collaborative network. The STEM scheme aims to encourage young people to study STEM subjects at school and in higher education. (It was created in 2004, as a response to a UK government report by Roberts, 2002 [3], which highlighted the need for high quality scientists and engineers to support the UK's productivity and innovation performance.)

A more recent addition to the network is a second UK professional body, the West Midlands branch of the Institution of Engineering and Technology (IET), who have provided support for recent events and have expressed interest in participating in future initiatives. The IET promises to be a good partner to the network, very actively supporting activities in schools in a broad range of disciplines, including technology, computing and engineering.

The participation of Warwick Institute of Education (WIE) provides an important link with future ICT teachers during their teacher training year on the WIE Postgraduate Certificate in Education (PGCE) course, a graduate route into teaching. During the 2007-8 schools competition ICT PGCE students on placements with local schools entered teams of pupils from their placement schools for the competition, and the winner of the Key Stage 4 section of the 2007-8 was a current PGCE student.

Effective contacts have been established with ICT teachers in local state schools, private schools and further education colleges as a result of the activities to date. To date the activities have focused on secondary school level (covering ages 11 to 19) rather than primary level.

2.3 Network Resources

Building and maintaining the trust and support of individuals and organizations that can contribute in different ways towards activities and events is an essential part of the network's function. The network has no funds of its own to draw upon, as it's a purely collaborative activity. Fortunately the network includes many willing volunteers. With no external source of funding, there are no preconceived goals or predetermined objectives to meet. If an activity is deemed to be of value and it sufficient funding can be found from collaborators, it will happen, otherwise not. The lack of direct funding has not been a problem, rather – it has promoted collaboration.

There are a number of generous organizations associated with the network who are prepared to provide resources, funding or sponsor the events. The two universities contribute towards costs, resources and use of facilities. BCS Headquarters has provided substantial funding and materials to support specific events. More recently IET has contributed to events and has expressed strong support for future activities. There has been financial or in-kind support from all local government organizations involved in the network. Contributions have also been made by commercial companies, either as direct funding or by donation of competition prizes. The local Coventry press plays a key part by helping to promote and publicize the events.

3   Events, Activities

The BCS (Coventry Branch) competition is a regular event, running every two years, with the current competition starting in September 09 and running through to April 2010, culminating in an awards ceremony for competition finalists in June 2010. The competition aims to encourage young people to use IT in creative communication applications and to support the development of a range of skills including teamwork, project planning and specific technology skills. One important aspect of the competition is its low threshold to participation. Schools teams can take part in the competition either by working together in class time as a class based activity, through an afterschool club, or by groups of motivated pupils in a school forming their own teams.

During the previous competition in 2007-8, both to support the competition, and to provide networking and discussion opportunities, a number of afterschool workshops for teachers were organized to run alongside the competition, mainly on topics suggested by teachers. In the light of the success of the workshops, a range of similar events is being planned during the current competition.

In order to ensure the focus of these events stay relevant to teachers, but don't duplicate other activities already aimed at ICT teachers, a regular monthly network meeting was set up between ICT Advisors from both Coventry and Warwickshire local education authorities (LEA), the local STEM coordinator (Science and Engineering Ambassadors), and the authors representing Coventry and Warwick Universities and BCS. At least one activity or event per term of interest to teachers. has been coordinated through these meetings. Various channels available to the meeting participants have been used to promote the events. The culmination of the interim year activities from the network was a one day conference in June 2009, drawing together representatives involved in secondary and tertiary education, and professional. The conference was utilized to launch and publicize the 2009-10 schools competition. Feedback from the conference participants identified a number of ideas which could be developed into potential initiatives for future collaboration.

4   Motivation

4.1   BCS (Coventry) Volunteers

In any IT related industry it can be difficult to keep up to date with latest developments and directions; teaching is no different, so activities which inform and give context to teachers are helpful. It is also useful for professionals working in the IT industry to have an awareness of how (and what) IT is taught in schools and universities. Building links between these groups is to the advantage of all.

Many individuals in the network demonstrate a high level of altruism and professional commitment by supporting their discipline and by contributing very specialist knowledge.

4.2 Tertiary Education

A significant motivator for universities to work with schools is to raise their institutional profile, aid recruitment of good students and promote wider participation in higher education. In the area of computer science there has been a particularly strong need recently to encourage more participation in computer science, following an international downturn in the number of applications to the subject between 2004 and 2009 as evidenced through UK research commissioned by the Council of Professors and Heads of Computing (CPHC) reported by Anna Round, who reports seriously declining application and admission levels for computing subjects against an upward trend overall in the UK [4]. David Geer in 2006 [5] provided a global view that highlighted the widening gap between the increasing requirements of industry, and declining numbers of graduates. The most recent applications figures for the two universities in Coventry appear to be showing a healthy reversal to this trend.

The recent decline affected both male and female applications to university. However it is widely known that there has been a long-standing and serious gender disparity in recruitment to computer science affecting many countries, which has been the subject of much research for over 20 years. In the UK Women into Computing (WiC) was founded to coordinate activities. WiC organized a series of international conferences to highlight the problem and to provide a forum to disseminate research in this area, from 1987 [6]. In her summary of the 1997 WiC Conference Janet Stack said "The need for action to increase the number of women applying to computing courses became apparent to universities when statistics showed that the percentage of women accepted by UK universities fell from 24% in 1980 to 10% in 1987" [7]. Activities involving school pupils and ICT teachers need to be designed to be inclusive to female pupils and to encourage females to seriously consider career options in computer science.

4.3 Professional Bodies

Part of the mission of all professional bodies is for public engagement and to promote the profession. The BCS is particularly interested in encouraging good practice relating to ICT in education, not least by offering IT skills certification for schools such as the European Computer Driving Licence (ECDL) [8] and more recently introducing the Digital Creator Award (Cre8tor) [9].

An article in the BCS journal IT Now by David Evans [10] published in January 2009 set out the corporate BCS view and provided a number of reasons why BCS members should engaging in activities to promote the profession to younger people. He highlighted that the academic computer science community is the most active sector, but that there are opportunities for all BCS members to become involved with awareness raising activities. The Coventry network is a perfect example of such a scheme, it involves a range of interested computing professionals engaging in community and schools activities in order to provide opportunities to develop skills and knowledge about ICT and careers.

A second UK professional body, the IET has interest in supporting ICT in schools and is actively contributing to the funding and support for the network's events and activities.

4.4 Sponsors and Other Beneficiaries

It has been the policy of the network not to make any charge to participants of the events, but instead to seek funding from various sources and organizations to cover costs, the aim is to remove barriers to participation, it is often easier to attend a free event, than to look at justifying provision of funding.

When organizing events there is normally an obligation to provide some publicity for the sponsors and the participants, either to encourage participation or to report the outcome after the event. Particularly where schools, teachers and pupils are involved the publicity can be considered part of the reward for participation, and local newspapers have provided support by running articles about events. The wider network provides scope for increasing and diversifying sponsorship and also the level and media for dissemination.

4.5 Secondary Education

A discussion on motivation within this context would not be complete without consideration of the potential beneficiaries of these initiatives, namely the schools, ICT teachers and their pupils. Schools are generally interested in participating in externally organized events as long as there is a clear benefit to the school. Staff development through free seminars, workshops and conferences could be considered of interest to schools. However if events in any way interfere with the operation of the school, perhaps necessitating hiring of substitute teaching cover, then these are barriers to participation.

The subject content in English schools is largely driven by the national matriculation system. The traditional academic route is GCSE ordinary level qualifications at level 2 (age 15-16) followed by advanced level qualifications at level 3 (age 17-18), although recently a minority of English schools have adopted the International Baccalaureate. There are also alternative vocational qualifications in the form of GNVQ and advanced GNVQ. However the vocational qualifications are being replaced by a system of new 14-18 Diplomas, which are gradually being introduced regionally according to subject focus.

Since 2006 there has been a general downward trend in school pupils taking ICT qualifications at level 2 in the UK, as can be seen from Table 1, which shows selected statistics from the whole of the UK taken from the Joint Council for Qualifications (JQC) web site [11]. The General National Vocational Qualifications (GNVQ) and short courses in ICT show a similar profile.

Table 1. UK Level 2 qualifications, 2005-9 adapted from JQC [11]

ALL UK	2005	%	2006	%	2007	%	2008	%	2009	%
GCSE ICT Males	58713	2.1	60888	2.1	55150	1.9	47561	1.7	40629	1.5
GCSE ICT Females	44689	1.5	48713	1.7	44506	1.5	38038	1.3	32890	1.2

At level 3 in English schools there is the choice of advanced level General Certificate of Education (A Levels) in either Computing or ICT. The Computing A-levels include some computer science content, but the ICT A-levels are less technical in content and largely focused on the ICT use and applications.   The candidate numbers on these courses demonstrate a similar decline, as can be seen from Table 2.

Table 2. UK Level 3 qualifications, 2005-9, adapted from JQC [11]

ALL UK A-level	2005	%	2006	%	2007	%	2008	%	2009	%
Computing Males	6426	1.8	5629	1.5	5035	1.4	4588	1.2	4256	1.1
Computing Females	816	0.2	604	0.1	575	0.1	480	0.1	454	0.1
ICT Males	8632	2.6	8658	2.4	8378	2.3	7607	2.0	7339	1.9
ICT Females	4599	1.2	4399	1.1	4986	1.1	4670	1.0	4609	1.0

Table 2 compares the popularity of Computing and ICT A-levels.   The statistics from both tables also demonstrate the serious gender imbalance in both ICT and Computing and the very low take-up of the more technical Computing A-level by female candidates.

5   Challenges

The UK Midlands area surrounding Coventry has complicated arrangements for local government and educational provision, cutting across jurisdiction demarked by several county and city boundaries. Each sub-region has autonomy over many aspects, including school holiday dates, age progression arrangements between schools (for example south Leicestershire has middle schools catering for ages 12-14). It can be difficult to keep abreast of all the circumstances and variations in the sector.

There can be problems making contact with teachers, identifying the most appropriate person or finding the best time and method to get in touch with them. The normal school schedule means that teachers are often not able to receive or respond to messages, either by telephone or email during the school day in term-time. During school holidays it is often difficult to make contact with teachers.

When organizing an event it is difficult to find an ideal date and time to suit teachers, as schools often have their own evening activities. ICT teachers in particular often have more general ICT support duties and demands on their time. Ideally ICT teachers should be motivated to join in activities which help them to update skills, generate ideas, develop new materials and generally help to improve the classroom experience for their pupils. However attending workshops may mean giving up personal time, and entering competitions could potentially generate a great deal of extra work. Such factors can be seen as disincentives to busy teachers. For all the reasons mentioned above, even when teachers identify an event or activity they would like to participate in, it may not be possible for them to do so.

The availability of IT and Computing in English schools is a major area of interest to this paper. There has not been a GCSE Computing qualification (level 2) available for some years. The take-up of ICT qualifications at level 2 initially increased significantly up to 2006, but since then has been to decreasing (Table 1). There has been a significant decline in the number of pupils taking Computing at level 3 as seen from Table 2 [11]. One of the qualification boards Edexcel no longer offers A-level Computing. The advent of league national tables for secondary schools has increased the pressure on schools to maximize their overall results to remain competitive. ICT qualifications at all levels are generally less technical than the equivalent Computing qualifications. It is clear that there could be more advantage in terms of higher grades achieved by schools offering ICT rather than computing. However the downward trend in ICT candidates a potential subject for further study.

The full reasons for the decline in interest in Computing and ICT qualifications are not clear, but two possible factors are maintaining subject currency, particularly in the technical skills, and other commitments of ICT teachers. The shortage of ICT specialists has resulted in ICT in some schools being taught by teachers with no formal qualification in the subject. Even specialist ICT subject teachers need to constantly re-skill to keep up to date in their field.

Exploration of two English GCE A-level syllabi for computing reveals rather dated content, including the paradigms and methodologies abandoned some time ago in universities, for example the OCR A-level Computing syllabus requires candidates to "produce and describe top/down/modular design" and use "a program flowchart", but the 2008 version does also make reference to Rapid Application Development and prototyping [12]. The new AQA Computing syllabus for implementation from 2009 is very similar to the OCR 2008 offering [13]. Such subject content may fail to convey to pupils the modernity and excitement of computer science as a career direction.

There are many other factors that could have influenced the profile of candidates to Computing and IT qualifications. Given the preference for ICT rather than Computing and the minimal technical content required in ICT, there cannot be any assumption or high expectation about the technical computer science skills of ICT teachers or their pupils.

The above limitations had to be considered when establishing a relevant and inclusive computing competition for local schools. It was important to set a task that was as demanding as possible, but still viewed by the pupils and teachers as achievable within the available time-frame. Even after the experience of two competitions, there remain difficulties convincing some of the computing professionals making up the competition schools committee and judging panel about expectations in skills and standards of entries.

Each of the network participants has their own agenda, constraints and reasons to be involved. The aims and objectives are normally complementary and mutually supportive, but there have been occasional misunderstandings, typically arising from lack of appreciation of different perspectives. Finally, there is need to recognize that the network operates on a voluntary basis, and its aims are to build long term relationships. It is important that individuals and groups ensure they participate at a level that they can comfortably sustain, and in ways that fit round other commitments.

6 Reflections

Evaluation following events provides indication of success and suggestions about improvement for the future. It is important that opinions continue to be sought and analysis carried out after every event, to ensure that problems any new ideas can be identified to ensure that future events are suitably enhanced.

With the variety of organizations involved clear communication becomes important and can be quite time consuming. To date, the network has relied on communications between individual members of the different groups, with decision making mainly being made though meetings. As relationships and trust between the different groups grow, it should be possible to move towards a more dynamic working model, and sub groups may form to focus on specific areas of interest.

Participation in events such as IT competitions is less likely if significant burden of responsibility is placed on the school. Instructions and guidance needs to be clear and accessible. It can be a great asset to an ICT teacher to have ready-made activities with external rewards and recognition that fit existing assessment requirements. However the notification and detail of activities must be appropriately timed and flexible to allow for schools to plan and organize, ideally before the start of the school year. In recognition of this from previous experience the 2009-10 BCS Coventry Branch Schools competition was launched in June 2009, circulated by email, with web-based guidance and support [14].

Network participants have expressed diverse opinions about the requirements for competitions for schools and there are great variations in expectations for outcomes. Notably there was a degree of disappointment from the judges in the standards achieved by entrants to the schools competitions submitted in 2006 and again in 2008. From the BCS professional body perspective one of the aims of the schools competition is to reward excellence. However there must also be a strong developmental aspect to such activities. The wider network is providing scope for clearer understanding of the reasons behind observed lower standards in some entries. These include limitations of resources and expertise in some schools, but also the ICT qualifications content and requirements are significant and affect all schools. This intelligence is providing the impetus to develop suitable guidance and support structures for teachers and their classes, with the aim of helping to drive up standards in the longer term.

The network is continuing to grow in both numbers and diversity, to the benefit of all concerned. The number of volunteers helping to organize the 2009-10 schools competition and associated events is at an all-time high. However it would be counterproductive to overburden individuals or make excessive demands on their time. The online collaboration tool "huddle" was utilized for communication, sharing of documentation and version control in the lead up to the launch of the 2009-10 schools competition. Although there were some initial difficulties using this tool, there are plans to continue with this as the primary asynchronous means of communication during the 2009-10 competition.

Although the focus for the network has been the staging of events and activities for teachers and pupils, and it is clear to see the benefits they derive from these activities, other benefits also exist. One additional benefit has been to raise awareness of the position of IT in secondary schools. Another important benefit has been the creation of links and collaborations between individuals in the network, creating informal interest groups. One of these groups focused on the MIT's Scratch programming language [15] and has supported its introduction to schools in the local area. Scratch is recognized as an attractive introduction to programming and is suitable for children at primary and secondary schools [16], appealing equally to boys and girls.

7 Future Plans

The network will thrive providing it continues to serve a useful purpose for participants. However the authors acknowledge that they would like to encourage more inclusivity and generate new ideas that will help to support the work of associates. This could be achieved through enhancing the collaborative framework, to better support different initiatives reflecting the interests of network members. More work needs to be done to bring this about. As identified earlier, there are several new directions to be explored. For example it may be useful to link our activities to other initiatives that have similar themes. This would provide the potential for establishing cross-regional or international networks for collaboration and could be an exciting prospect for the schools involved in the Coventry regional network.

The introduction of the new 14-19 IT diploma locally from 2010 is likely to drive changes in schools' teaching and learning methods and require collaboration with industry and possibly academia to add context and relevance to the subject content. The network would provide a good vehicle for helping schools to develop the required links for the IT diplomas.

So far the activities have largely concentrated on links with secondary education. However there is evidence that primary schools would welcome some involvement or support in the area of ICT activities or projects. There is potential for involving primary schools in current initiatives or creating new activities designed specifically for younger children.

One possible future direction for the network to grow would be to link with the student volunteering initiatives at Coventry and Warwick Universities. Both Universities have programs that promote students volunteering to participate in local community and education activities. At Warwick one of the authors has piloted a scheme called Technology Volunteers, which runs as a part of the existing student volunteer schemes (Warwick Volunteers). This linked students with technology based skills with specific projects and afterschool clubs in local schools. In the coming academic year, this scheme will also link to the STEM initiative, with students becoming STEM Ambassadors (Science Technology, Engineering and Mathematics Ambassadors – national voluntary scheme setup in response to the Roberts 2002 report [3]). This scheme exploits the advantage that students are closer in age to pupils therefore can have more empathy and influence in their development.

The UK government has a mission for encouraging universities to accept more disadvantaged and "first in family" students to degree programmes, under the banner of "Widening Participation". Various initiatives exist to raise expectations in school children, improving outcomes and standards of qualifications of target students through direct involvement in schools and through positive interventions. This would be a very rewarding and worthwhile extension to the activities of the network, possibly making use of the student mentoring schemes already in place.

Recent workshops organized for ICT teachers have provided support, encouragement and aimed to generate confidence in the use of novel software tools in the classroom. The choice of subjects for the workshops were informed by the teachers themselves and as a result these have proved very popular. This activity is important because it has the potential to enhance the classroom experience for the current generation of school pupils, and skills in this area are needed for industry and commerce. The intention in future is to diversify the types of workshop offered and the range of audiences for which they cater, including non-ICT teachers, and primary school staff while liaising with LEA advisors to co-ordinate and focus activities.

The brief discussion and speculation about the profile of candidates for IT and Computing qualifications in schools has highlighted an interesting area for future research. The network provides an excellent starting point for research to establish the root causes for the phenomenon.

The different groups and bodies that participate in the network are experts in their own domain, and the hope is that through regular meetings, discussions and debate, all become more informed about different aspects of the profession, inspiring the next generation of computer scientists.

The creation of this paper has necessitated introspection through examination and comparison of many aspects of the network, including motives, practice and aspirations. As a result several hitherto unrecognized issues have been identified for future action, which should further strengthen the purpose and working practices of the network for the benefit of all concerned.

It remains to be seen how the network will develop. It is hoped that this account of the formation and purpose of the network will provide a useful case study for others to consider adopting as a model for collaborative practice.

References

1. Kozma, R., McGhee, R.: ICT and Innovative classroom practices. In: Kozma (ed.) Technology, Innovation and Education Change: a Global Perspective, pp. 43–80 (2003)
2. Preston, C., Mellors-Bourne, R.: Education, Education, Education. IT Now 50(1), 34 (2008)
3. Roberts, G.: SET for success: the supply of people with science, technology, engineering and mathematics skills. HM Government Report (2002), http://www.hm-treasury.gov.uk/ent_res_roberts.htm
4. Round, A.: The Decline in Student Applications to Computer Science and IT Degree Courses in UK Universities (2006), report on research commissioned by CPHC (accessed 10th, August 2009), http://www.cra.org/Activities/snowbird/2006/earnshaw.round.pdf

5. Geer, D.: Software Developer Profession Expanding. IEEE Software 23(2), 112–115 (2006)
6. Lovegrove, G., Segal, B.: Women into Computing (Selected Papers 1988-90). Springer, Heidelberg (1990)
7. Stack, J.: Women into Computing. ACM, New York (1997)
8. British Computer Society, European Computer Driving Licence, http://www.bcs.org/ecdl (accessed 6th August 2009)
9. British Computer Society, Digital Creator Award (Digital Cre8tor), http://www.bcs.org/cre8or (accessed 12th August 2009)
10. Evans, D.: Children and Careers in IT. BCS IT Now, Part 1, 51, 20–21 (2009), http://www.bcs.org/server.php?show=ConWebDoc.24152 (accessed 19th August 2009)
11. Joint Council for Qualifications, UK national results, http://www.jcq.org.uk (accessed 19th August 2009)
12. OCR Qualifications website, http://www.ocr.org.uk (accessed 19th August 2009)
13. AQA Qualifications website, http://www.aqa.org.uk (accessed 19th August 2009)
14. BCS Coventry Branch Schools Competition Website, http://coventry.bcs.org/competition.php
15. MIT's Scratch programming language, http://www.scratch.mit.edu (accessed 12th August 2009)
16. Monroy-Hernández, A., Resnick, M.: Empowering kids to create and share programmable media. ACM Interactions, 50–53 (March-April 2008)

About the Authors

Irene Glendinning is Academic Manager for Student Experience in the Faculty of Engineering and Computing at Coventry University, with over 30 years experience as a Computer Scientist in industry and education. Margaret Low is a Senior Teaching Fellow in WMG (Warwick Manufacturing Group) at the University of Warwick and has been teaching software development since 1988.

Teaching Public-Key Cryptography in School*

Lucia Keller, Dennis Komm, Giovanni Serafini,
Andreas Sprock, and Björn Steffen

Department of Computer Science, ETH Zurich, Switzerland
{lucia.keller,dennis.komm,giovanni.serafini,andreas.sprock,
bjoern.steffen}@inf.ethz.ch

Abstract. These days, *public-key cryptography* is indispensable to ensure both confidentiality and authenticity in numerous applications which comprise securely communicating via mobile phone or email or digitally signing documents.

For all public-key systems, such as RSA, mathematically challenging and technically involved methods are employed which are often above the level of secondary school students as they employ deep results from algebra. Following an approach suggested in 2003 by Tim Bell et al. in *Computers and Education, volume 40, number 3*, we deal with the question of how to teach young students the main concepts, issues, and solutions of public-key systems without being forced to also teach rather complicated theorems of number theory beforehand.

1   Introduction

Cryptography is one among the topics in computer science that raise the interest of most students of all ages. Naturally, it allows the teacher to create interesting lessons including experiments and games. Complementing the well-known classical cryptosystems, such as Caesar and Skytale [1], which are rather easy to understand, in this paper we want to focus on communicating the ideas of mechanisms that are actually used today. Some of the most widely used and popular cryptosystems (e. g., RSA, El-Gamal, . . .) are asymmetric methods based on the idea of *one-way functions with trap-doors*. For many students it is fascinating that there are functions in which computation in one way is easy and the computation of the inverse function is much harder. At this point, it is the challenge of didactics to introduce the key issues behind this concept in a fashion as simple as possible. To explain the intuitive idea, there is a famous example given by Salomaa [14] using a phone book. Since nowadays this example is not really applicable, we use a more sophisticated model for our lectures. We took the idea from "CS unplugged" [2] and attempted to duplicate the positive results concerning the progress of students [3] similar to the work of Nishida et al. in Japan [12]. In contrast to their work, we only focused on one specific topic from "CS unplugged".

To describe our concept in greater detail, we first need to give a quick overview of the most important facts about public-key cryptography. The need for

---

* This work was partially supported by the Hasler Foundation, project "Kantonales Fortbildungszentrum Informatikunterricht", and by FILEP grant 351 of ETH Zurich.

J. Hromkovič, R. Královič, and J. Vahrenhold (Eds.): ISSEP 2010, LNCS 5941, pp. 112–123, 2010.

asymmetric cryptosystems arises from the following problem: when using symmetric-key cryptosystems, at least two parties *share* a secret (i.e., a key) which they use to encrypt and decrypt messages from plaintext to ciphertext and the other way around. In the real world, however, a serious problem arises when trying to share the key, because clearly, in general, the two entities (in this scenario, say two persons Alice and Bob) do not have a secure channel to communicate yet. Many strategies have been proposed to solve this problem, among which one of the most famous certainly is the Diffie-Hellman key exchange protocol [8]. However, on the downside, using this protocol still requires the creation of approximately n^2 keys if n parties want to securely communicate with each other. In 1978, one of the first and (up to today) most widely used public-key cryptosystem, namely RSA, was introduced [13]. Here, it suffices to create n key pairs for a scenario as above. For more details, we refer to the standard literature [1, 7].

The main idea behind public-key (or asymmetric) cryptosystems is the following: one entity has (in contrast to symmetric cryptosystems) a pair of keys which are called the *private* key and the *public* key. These two parts of the key pair are always related in some mathematical sense. As for using them, the owner of such a key pair may publish her public key, but it is crucial that she keeps the private key only for herself. Let (sk, pk) be such a key pair where sk is Alice's secret private key and pk is the corresponding public key. If a second person Bob wants to securely send Alice a message M, he computes $C = encrypt(M, pk)$ where *encrypt* denotes the so-called *encryption function* which is also publicly known (see Fig. 1). This function is a one-way function with a trap-door. In other words, the trap-door allows for the creation of the secret key sk which in turn enables Alice to easily invert the encryption function. We call C the ciphertext. Obtaining M from C can be done easily using the (publicly known) decryption function *decrypt* and Alice's private key (sk). On the other hand, it is much harder to decrypt without having any knowledge of the private key. As already mentioned, the great advantage of this approach is that no secure key exchange is necessary before a message is transmitted.

Since public-key cryptography is commonly used in many applications and settings, it is very appropriate and promising to teach cryptography. However, since the low level ideas and calculations require a lot of mathematical background, it is very hard to discuss cryptosystems like RSA with students of secondary schools.

We therefore use a very simple and straightforward theoretical cryptosystem from "CS unplugged" by Tim Bell et al. [2] that only involves some very easy

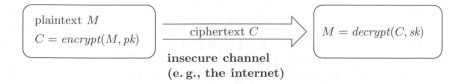

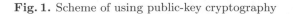

Fig. 1. Scheme of using public-key cryptography

mathematical ideas, but highlights the important principles of public-key cryptography very nicely.

The paper is organized as follows. In Section 2, we describe the cryptosystem used and explain its theoretical background. After that, we explain how to use this approach to teach students the ideas of public-key cryptography and also describe our experiences in Section 3 and 4. Section 5 discusses a concrete lesson held in July of 2009 at a local secondary school in Switzerland. We conclude with a reflection of the main ideas and open questions in Section 6.

2   Technical Details for the Teachers

The cryptosystem used is based on a graph-theoretical problem which we describe in the following. First, we need some basic notations and definitions.

For $n \in \mathbb{N} \setminus \{0\}$, let $V = \{v_1, v_2, \ldots, v_n\}$ denote a set of *vertices* and $E \subseteq \{\{v_i, v_j\} \mid i, j \in \{1, 2, \ldots, n\}, i \neq j\}$ a set of *edges*. We then call $G = (V, E)$ a *graph* with n vertices and $|E|$ edges. Furthermore, if for two vertices v_i and v_j we have $\{v_i, v_j\} \in E$, we call v_i and v_j *adjacent* (i.e., connected by an edge). For $i \in \{1, 2, \ldots, n\}$, let $weight : V \to \mathbb{Z}, v_i \mapsto w_i$ be a vertex *weight function* where the integer w_i is called the weight of vertex v_i. We call the pair $(G, weight)$ a weighted graph. For our investigations we need the following definition.

Definition 1 (Dominating set). *Let $G = (V, E)$ be a graph. A dominating set – DS for short – is a set of vertices $V_{\text{DS}} \subseteq V$ such that, for every vertex $v' \in V \setminus V_{\text{DS}}$, there exists a vertex $v \in V_{\text{DS}}$ with $\{v', v\} \in E$.*

A special variant of a DS has the additional property that, for every vertex from $V \setminus V_{\text{DS}}$, there exists exactly *one vertex in V_{DS} such that $\{v', v\} \in E$ and that no two vertices from V_{DS} are adjacent. We call this special case an* exact dominating set *– EDS for short – and the corresponding set V_{EDS}.*

In the following, we construct a graph with an EDS. The graph itself is used as the public key and the set V_{EDS} as the secret key. Note that determining whether a graph has an EDS is an $\mathcal{NP}$-complete problem [6]. Thus, finding an EDS in a graph which is known to have an EDS is $\mathcal{NP}$-hard, too. Again, consider two persons Alice and Bob and assume that Alice wants to securely send a message M to Bob. Bob therefore needs to create a key pair (sk, pk) to enable them to encrypt and decrypt messages and therefore he takes the following steps.

a.1 Choose two arbitrary natural numbers n_{EDS} and n_{dom} and let $V_{\text{EDS}} = \{v_1, v_2, \ldots, v_{n_{\text{EDS}}}\}$ and $V_{\text{dom}} = \{v'_1, v'_2, \ldots, v'_{n_{\text{dom}}}\}$ be two pairwise distinct sets of vertices.

a.2 Set $E = \emptyset$. Then, for *every* vertex $v' \in V_{\text{dom}}$, choose *exactly* one arbitrary vertex $v \in V_{\text{EDS}}$ and set $E = E \cup \{\{v, v'\}\}$, i.e., connect every vertex from V_{dom} to exactly one vertex from V_{EDS}.

a.3 Choose an arbitrary number of pairs (v'_i, v'_j) of vertices from V_{dom} and set $E = E \cup \{\{v'_i, v'_j\}\}$, i.e., connect v'_i to v'_j.

a.4 Set $V = V_{\text{EDS}} \cup V_{\text{dom}}$ and $G = (V, E)$.

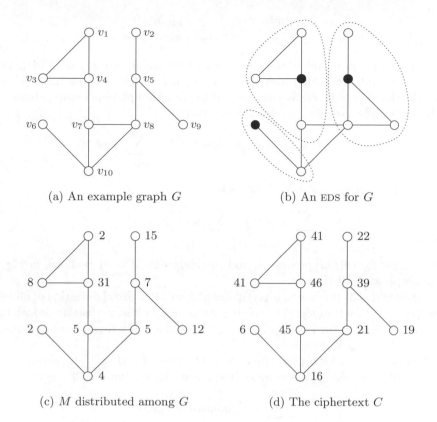

(a) An example graph G (b) An EDS for G

(c) M distributed among G (d) The ciphertext C

Fig. 2. An example of a graph G for encrypting the plaintext $M = 91 = 2 + 15 + 8 + 31 + 7 + 2 + 5 + 5 + 12 + 4$

As already mentioned, the graph $G = (V, E)$ is Bob's public key pk (as shown in Fig. 2 (a)). Obviously, the set V_{EDS} forms an EDS which is immediately clear by the way the edges of G are constructed in a.2 (as shown in Fig. 2 (b)). Bob keeps V_{EDS} as the private key sk for himself and publishes G.

Now suppose that Alice wants to encrypt the plaintext message M using the public key of Bob. Let $M \in \mathbb{N} \setminus \{0\}$ be a natural number.[1] She acts as follows.

b.1 Write M as the sum $\sum_{i=1}^{n} w_i$ where w_i are integers which are randomly chosen (except the last one).

b.2 Then define a weight function $plain(v_i) = w_i$ for every $v_i \in V$. The "plain" graph $(G, plain)$ is shown in Fig. 2 (c).

b.3 For every vertex $v \in V$, let neighbor$(v) = \{v' \mid \{v, v'\} \in E\} \cup \{v\}$ be the set of all neighbors of v in G, i.e., all vertices which are adjacent to v and v itself. We then define the function $ciph : V \to \mathbb{Z}$ as

---

[1] Clearly, we may represent any text by natural numbers using, e.g., ASCII.

$$ciph(v) = \sum_{v' \in \text{neighbor}(v)} plain(v').$$

The weight function $ciph$ is the ciphertext of M (as shown in Fig. 2 (d)). Obviously, $encrypt$ is the algorithm described by b.1 to b.3. The plaintext M, on the other hand, can be easily calculated if the function $plain$ is known. However, Bob may simply use sk to receive the plaintext with

$$decrypt(C, sk) = decrypt(ciph, V_{\text{EDS}}) = \sum_{v \in V_{\text{EDS}}} ciph(v).$$

Since V_{EDS} forms an EDS in V, it holds that

$$\sum_{v \in V_{\text{EDS}}} ciph(v) = \sum_{v \in V} plain(v) = M.$$

An example for this cryptosystem and the plaintext $M = 91$ is shown in Fig. 2 for a graph with 10 vertices.

However, if sk is unknown, it is still possible to calculate the weights of all vertices (i. e., the function $plain$) by solving a system of n linear equations which can be done in $\mathcal{O}(n^3)$ which clearly is extensively more time[2] than the computation of M takes if V_{EDS} is known.[3]

More particular, for a given ciphertext, the message $M = \sum_{v \in V} plain(v)$ can be decrypted by solving a system of linear equations given for all $v \in V$:

$$\sum_{v' \in \text{neighbor}(v)} plain(v') = ciph(v)$$

For the example shown in Fig. 2 (d), we obtain the following system of linear equations.

$$plain(v_1) + plain(v_3) + plain(v_4) = ciph(v_1) = 41$$
$$plain(v_2) + plain(v_5) = ciph(v_2) = 22$$
$$plain(v_1) + plain(v_3) + plain(v_4) = ciph(v_3) = 41$$
$$\vdots \qquad\qquad \vdots \quad = \quad \vdots$$
$$plain(v_6) + plain(v_7) + plain(v_8) + plain(v_{10}) = ciph(v_{10}) = 16$$

Please note that such a system might have one or infinitely many solutions, but the sum is still unique. A reasonable definition of a secure cryptosystem for teaching purposes is:

---

[2] Note that the fastest known algorithm can solve a system of n linear equations theoretically in $\mathcal{O}(n^{2.367})$ [5].

[3] That means that we still need to solve an $\mathcal{NP}$-hard problem to find sk, but for a given public key and ciphertext $(G, ciph)$, the plaintext M can be computed in polynomial time.

> A cryptosystem is secure if there does not exist any efficient algorithm
> that decrypts the cipher text without knowing the secret key used, but
> knowing the way in which the cryptosystem works. [11]

This implies that the cryptosystem introduced is not *secure* in this sense. However, it is still very adequate to explain asymmetric-key cryptography since we
can easily show our students the big discrepancy concerning the effort between
decrypting with and without knowing sk.

3 How We Teach Public-Key Cryptography

In what follows, we give an exemplary way of how to teach students the ideas
and concepts of public-key cryptography using the cryptosystem introduced in
Section 2.

Usually the students had some previous classes, where we taught them the
basic concepts of classical (symmetric) cryptography.

3.1 Teaching Goals

We define the following teaching goals – they describe the minimal knowledge
the students should achieve after completing the teaching unit:

1. Students understand why in *public-key cryptography* two different keys are
   needed for encryption and decryption. One is public, the second one must
   be kept secret and is only known to the owner (is private).
2. Students are able to explain that everyone is able to encrypt the plaintext
   with the public key, but only the owner of the related private key has the
   means to decrypt the ciphertext.
3. Students are able to depict the decryption process as an easy task when a
   secret information is known and as a practically infeasible one if this information is not available.
4. Students are able to encrypt and decrypt messages correctly with the presented cryptosystem.

3.2 Introducing the Concept of Public-Key Cryptography

First, we motivate the need for public-key cryptography by showing that the
key management overhead of symmetric cryptography is high. We point out the
weaknesses of symmetric cryptosystems and propose the concept of having two
keys (a private and a public one) which clearly overcomes the disadvantages of
symmetric cryptosystems. We informally introduce one-way functions with trap-
doors and then ask the students whether they are familiar with anything like it
in everyday life.

As a first candidate of such a function, we describe the public-key cryptosystem that uses a phone book [14]. Given a phone book (the public key), we can
encrypt a letter by choosing at random a name starting with this letter and then

send the corresponding phone number instead. Clearly it is easy to find a name starting with a given letter. On the other hand, it is hard to find the name which belongs to a given phone number. But if the recipient has a special phone book (the private key) which is ordered according to phone numbers, then she is able to efficiently decrypt the message.

3.3   Teaching the Graph-Theoretical Cryptosystem

To give a more qualified cryptosystem we now explain the students the graph-theoretical public-key cryptosystem discussed in Section 2.

To do this, we first informally introduce them to graphs. For the cryptosystem it is only important that the students know the notions of *vertices*, *edges* and the concept of a *neighborhood*. Note that we do not describe what an exact dominating set is, yet.

We now show the students an example graph G with n vertices that has an EDS which is, of course, only known to us. This graph represents a public key and we explain how to encrypt an integer (message) with this graph (as shown in Fig. 2):

Step 1. We draw the graph G on the blackboard to encrypt the message M.

Step 2. We write M as a sum of n integer summands.

Step 3. We write each of these numbers next to exactly one of the n vertices (using some color, e. g., green) and ask the students if this is already a secure ciphertext.

Step 4. We then show the students how to obtain the ciphertext by adding up the numbers in the neighborhood of a vertex and writing the sum down in some other color (e. g., red) next to the corresponding vertex. After calculating all red numbers, we clean out the green ones and say that the graph G with the red numbers is our ciphertext.

Asking the same question as above, the students realize that it is now not easy anymore to derive the plaintext given only the red numbers.

After all students know how to encrypt a message using this cryptosystem we may carry out a little contest within the class. We form teams of two students each. Every team is given two copies of a graph different from G on two pieces of paper and they have to encrypt a message which they have chosen for themselves. Now, the students can write the summands of the plaintext on one graph, each number next to the according vertex. After calculating the numbers for the ciphertext, they then write these numbers on the second piece.

When all teams are finished, we collect their ciphertexts and give each team the same ciphertext that we have created beforehand. Their task is to find the plaintext of the given ciphertext.

In the meantime, we demonstrate how easy the recipient can decrypt the message. We do this by decrypting the messages of the teams very fast and show them that we really found the corresponding plaintext. The students are usually very impressed how fast we can do that, as they have not been able to decrypt the given ciphertext, yet.

After that, EDS are introduced on the board (usually, the students have gotten an idea of how the secret key might work during the competition).

The next step is to demonstrate that it is a hard task to find an exact dominating set in a graph. Achieving this is rather easy. We simply hand out a drawing of a big graph ($n \geq 30$) and ask the students to find its EDS.

Afterwards, we need to show that, on the other hand, it is an easy task to design a graph which has an EDS which is only known to the creator of the graph. Therefore, we draw a set of vertices to the board and mark them (again, we might use some special color) as *dominating vertices*. After that, we draw another set of *dominated vertices* in a different color. We then draw exactly one edge from each dominated vertex to one dominating vertex. Finally, we add various edges between the dominated vertices. It is obvious that the dominating vertices now really form an EDS for this graph. After this is also clear to the students, we color all vertices with the same color to "cover up our tracks".

What should follow is a discussion on how difficult it is to decrypt the plaintext with and without knowing the EDS. At this point, the students themselves probably discovered the alternative way to decrypt the message (i. e., a system of linear equations as described in Section 2). However, if not, we make them familiar with this idea and write down such a system for the graph on the board. Our students are then asked to solve it themselves. They will immediately realize that this way is a lot harder than using the EDS.

The investigation of the running time of algorithms for solving a system of linear equations may be as detailed as the student's knowledge allows it to be. In any case it is very important to show and to compare the two function graphs so that the difference between decrypting with and without the secret becomes clear, i. e., the linear function $f(n) = n$ for the decrypting knowing the EDS (we simply need to do less than n additions) and the function $g(n) = n^3$ for decrypting via solving a system of linear equations.

4 General Experiences

We have been able to gather a multitude of experiences by applying our teaching concept at various Swiss secondary schools which we describe in the following.

4.1 Introducing the Concept of Public-Key Cryptography

The students will certainly agree that the method in the example of the phone book makes sense, but they are not very convinced that this method works in practice, since, nowadays, with the help of digital phone books, anyone is able to find the name to a given phone number efficiently. We also have to exclude the option of calling the number and asking for the name.

This leads to the observation that this cryptosystem is rather artificial. But none the less, the example is very qualified to introduce the basic concepts of public-key cryptography and it gives a first idea of the difference between decrypting with knowing the secret key and decrypting without knowing it.

4.2   Teaching the Graph-Theoretical Cryptosystem

This example is a lot more serious and realistic and it has the big advantage that we do not have to restrict the students in the same drastic way as for the first candidate.

Of course, most students immediately realize that the message in step 3 is not encrypted at all, since one can easily add up the green numbers. But this question prepares the students for the following step and keeps their attention on the distribution of the numbers.

The students enjoy the competition and are highly motivated to find the secret of the decryption. In many classes that we taught, at least some groups were successful, only a few needed hints.[4] They also get the feeling that this cryptosystem is more "serious" than using phone books, because now they see that it is not immediately clear how to find the plaintext.

At this point, some students discovered that they can get M using a system of linear equations. In this case, we simply told them that this still takes very long and there is a way which is a lot faster. We then postponed a detailed discussion of the running time of both methods to a point when the second technique (i. e., using the EDS) is also discovered.

5   An Example Lesson at a Secondary School

In this section, we informally describe an implementation of the presented method at a secondary school in Zurich at the beginning of July 2009. Classes other than languages and literature are taken either in German or in Italian [9].

The Swiss secondary school system is extremely heterogeneous. A common Swiss-wide frame of regulations [4] is implemented in many slightly different ways by each of the 26 cantons.

The secondary school involved focuses on preparing students for academic studies in the field of arts. Since computer science was no official subject at the time of our lesson, the students did not have any prior knowledge in algorithmics and programming. More specifically, in contrast to many other classes which, at this point, already knew basic things like classical symmetric ciphers, this was the students' first lesson on cryptography.

One of the authors has a teaching appointment for mathematics at this secondary school and tried the presented method in one of his classes. Mathematics lessons during the first two secondary school years are given in small classes consisting of up to 12 students. Students at the end of the second year are able to deal with basic algebra and Euclidean geometry, linear and quadratic equations, trigonometry as well as systems of linear equations.

The chosen class consists of students who are between 16 and 18 years old. The lesson was carried out as a game during the last lesson of the school year just after the students received their grades.

---

[4] It suffices to say that only a linear number of operations are necessary, e. g., "Five operations are enough".

5.1   Experimental Setting

Although we did not implement a formal empirical study, some typical methodical constraints were fulfilled during our test lecture: we decided to assess the available specific prior knowledge in cryptography (especially in public-key cryptography), then to carry out the lecture, and to assess the students knowledge again. We want to focus on measuring the students' proficiency in the basic principles of public-key cryptography after completing the teaching unit described.

The assessments were carried out anonymously. Each student received a personal tag. We gave the lecture in a compact (no longer than 45 minutes), concentrated and unstressed way.

5.2   Overview of the Lesson

The lesson was designed as simple as possible. We chose examples and metaphors based on the everyday life of students inside and outside school. The teaching goals correspond to the ones we described in Section 3.1.

Pre- and Post-Test. Both the pre- and the post-test were conducted based on the same form containing the following 5 questions:

1. The concept of *symmetric cryptography* means ...

    ☐ that in order to encrypt and subsequently decrypt a message, two different pieces of information ("keys") are necessary.
    ☐ that in order to encrypt and subsequently decrypt a message, only one (secret) information is necessary.
    ☐ I don't know the concept of *symmetric cryptography*.

2. The concept of *asymmetric cryptography* means ...

    ☐ that in order to encrypt and subsequently decrypt a message, two different pieces of information ("keys") are necessary.
    ☐ that in order to encrypt and subsequently decrypt a message, only one (secret) information is necessary.
    ☐ I don't know the concept of *asymmetric cryptography*.

3. In *asymmetric cryptography* ...

    ☐ everyone is able to encrypt a message using the so-called public key.
    ☐ everyone is able to encrypt a message using the so-called private key.
    ☐ I don't know the concept of *asymmetric cryptography*.

4. In *asymmetric cryptography* ...

    ☐ everyone is able to decrypt an encrypted message rapidly using the so-called public key.
    ☐ only the authorized recipient is able to decrypt an encrypted message rapidly using the so-called private key.
    ☐ I don't know the concept of *asymmetric cryptography*.

5. In *asymmetric cryptography* ...

    ☐ the decryption of an encrypted message is a very short process, if we
    know a secret information, otherwise it is a very long one. Without this
    secret information it is practically impossible to decrypt the encrypted
    text.
    ☐ the decryption of an encrypted message is a very long process, indepen-
    dent of whether you know a secret information or not. For everyone, it
    is practically infeasible to decrypt the encrypted message.
    ☐ the decryption of an encrypted message is always a very short process,
    independent of whether you know a secret information or not. Everyone
    is able to decrypt the encrypted message without any effort.
    ☐ I don't know the concept of *asymmetric cryptography*.

Protocol of the Lesson. We started the lesson with a short talk about security in the internet and mentioning the need for security when, e. g., buying
books online. The need for confidentiality while communicating was illustrated
on the basis of the situation in which two students wish to exchange encrypted
messages. Symmetric and asymmetric cryptography issues and terms were introduced on the basis of the lock metaphor (where the public key is a padlock
and the corresponding private key is a key that opens it).

In the following phase, we introduced the cryptosystem illustrated in the
prior sections. The students were really impressed by the topics and by the way
how messages are encrypted or decrypted. The challenging task to find out the
trick, permitting to decrypt a ciphertext rapidly (as the teacher did) was really
motivating for the class. Here, no one suggested a solution based on solving a
system of linear equations. One student depicted an approach similar to the
expected solution (i. e., using the EDS), without being able to explain the details
of her strategy.

The fast decryption method was shown to the class and explained carefully.
After that, the post-test was done showing that almost all students reached the
teaching goals and were able to talk about the high-level ideas of public-key
cryptography as intended.

6   Conclusion

In this paper, we discussed a simple way of introducing the basic principles of
public-key cryptography to students of secondary schools. A more detailed discussion of the cryptosystem will be part of an upcoming book [10]. We gave concrete
suggestions how to use the theoretical cryptosystem designed by Bell et al. [3]
to teach students of young age public-key cryptography in an entertaining way.
Additionally, we discussed our experiences while teaching these ideas. Finally, we
highlighted a specific lesson at a school in Switzerland. The lesson we described
was not planned as a formal experiment. However, it was possible to verify
that, after completing the unit, the majority of the students mastered the goals
we set.

It therefore remains open to formalize this experiment and test this method on a larger set of students. However, the results of this small class match our general experiences when teaching cryptography.

References

1. Bauer, F.L.: Decrypted Secrets: Methods and Maxims of Cryptology, 4th edn. Springer, Secaucus (2006)
2. Bell, T., Fellows, M., Witten, I.H.: Computer Science Unplugged - Off-line activities and games for all ages (1999), www.csunplugged.org (last accessed: October 22, 2009)
3. Bell, T., Thimbleby, H., Fellows, M., Witten, I., Koblitz, N., Powell, M.: Explaining cryptographic systems. Computers & Education 40(3), 199–215 (2003)
4. Bundesrat and EDK. Verordnung des Bundesrates/Reglement der EDK über die Anerkennung von gymnasialen Maturitätsausweisen (MAR) (1995), http://www.sbf.admin.ch/evamar/reglemente/VO_MAR_1995_d.pdf (last accessed: October 22, 2009)
5. Coppersmith, D., Winograd, S.: Matrix multiplication via arithmetic progressions. In: Proc. of the Nineteenth Annual ACM Symposium on Theory of Computing (STOC 1987), pp. 1–6. ACM, New York (1987)
6. Cull, P.: Perfect codes on graphs. In: Proc. of the 1997 International Symposium on Information Theory, p. 452 (1997)
7. Delfs, H., Knebl, H.: Introduction to Cryptography: Principles and Applications. Springer, Heidelberg (2002)
8. Diffie, W., Hellman, M.E.: New directions in cryptography. IEEE Transactions on Information Theory IT-22(6), 644–654 (1976)
9. Elmiger, D.: Die zweisprachige Maturität in der Schweiz (2008), www.sbf.admin.ch/htm/dokumentation/publikationen/bildung/bilingue_matur_de.pdf (last accessed: October 22, 2009)
10. Freiermuth, K., Hromkovič, J., Keller, L., Steffen, B.: Kryptologie, Lehrbuch Informatik. Vieweg+Teubner (to appear, 2009)
11. Hromkovič, J.: Algorithmic Adventures. Springer, Berlin (2009)
12. Nishida, T., Idosaka, Y., Hofuku, Y., Kanemune, S., Kuno, Y.: New methodology of information education with "computer science unplugged". In: Mittermeir, R.T., Sysło, M.M. (eds.) ISSEP 2008. LNCS, vol. 5090, pp. 241–252. Springer, Heidelberg (2008)
13. Rivest, R.L., Shamir, A., Adleman, L.M.: A method for obtaining digital signatures and public-key cryptosystems. Communications of the ACM 21(2), 120–126 (1978)
14. Salomaa, A.: Public-Key Cryptography. Springer, Berlin (1996)

Towards a Methodical Approach for an Empirically Proofed Competency Model

Johannes Magenheim, Wolfgang Nelles, Thomas Rhode, and Niclas Schaper

University of Paderborn
name.surname@uni-paderborn.de

Abstract. This article aims to outline the methodical proceeding within the project MoKoM funded by the German Research Foundation (DFG) to refine a theoretically derived competency model comprising competencies in the domain of informatics modelling and system comprehension. In this context the first part of this article deals with theoretical aspects of conducting expert interviews and outlines an exemplary analysis of three interview transcripts. It will be illustrated how to empirically refine one of the most important dimensions of the competency model by taking the results of a qualitative content analysis into account. Finally, the illustrated methodical proceeding is reflected and discussed.

Keywords: Empirical Methods, Research in Informatics Education, Curriculum Issues, Communication Skills, Competency Model, Educational Standards, Qualitative Content Analysis.

1 Motivation

This article intends to describe the methodical proceeding within the project MoKoM funded by the German Research Foundation (DFG). Generally, this project persues three goals: (1) *developing an empirically proofed competency model* concerning informatics modelling and system comprehension. With regard to the output orientation of the German school system [3], [7] this model can be understood as a basis for the development of educational standards for these specified domains of informatics secondary education. In order to refine our theoretically derived approach of a competency model, the first part of this article describes how expert interviews were conducted and on which theoretical basis the interviews were planned. Subsequently, the main part of the article will demonstrate an exemplary analysis of three interviews with regard to one (of altogether four) competency dimension. Here it is shown how to refine this dimension by taking into account the results of the qualitative content analysis of the interview transcripts. This dimension is named *basic competencies* and constitutes one the two most important dimensions. The two other goals of MoKoM are (2) *to develop and to test instruments* which are appropriate for competency measurement and in a further step (3) *to design and to evaluate effective learning environments* for competency development aiming at the two domains "informatics modelling" and "system comprehension". The development of an

J. Hromkovič, R. Královič, and J. Vahrenhold (Eds.): ISSEP 2010, LNCS 5941, pp. 124–135, 2010.

empirically proofed competency model becomes even more worthwhile because such a model is prerequisite for achieving the two last mentioned goals and to provide students with competency oriented informatics education.

2   Theoretically Derived Competency Model

Up till now there exists no adequate competency model for the domains "informatics modelling" and "system comprehension". Generally, competency models give a pragmatic response to educational debates and furthermore they provide orientation concerning professional teaching. They distinguish between different competency dimensions within a domain and they point out different levels of competency.

Our theoretically derived competency model consists of a total of four dimensions which are subdivided in further components. The dimensions are *Basic competencies, Perspectives towards an informatics system, Complexity of an informatics system* and *Non-cognitive competencies* [5]. The first dimension comprises knowledge and skill elements concerning *system development, system comprehension* and *system application*. The application of these competency components is furthermore graduated according to following levels: (1) Knowledge, (2) Transfer and (3) Evaluation. The second dimension consists of outer and inner perspectives towards an informatics system such as *usability, algorithms and data structures, graphical notation techniques*. This dimension is graduated as well by means of the same taxonomy. The third dimension deals with the demands on handling complexity of informatics systems. Among others, *Degrees of interactivity, Degrees of networking* are components of this dimensions. The fourth dimension is subdivided in three components: *attitudes, social-communicative skills* and *motivational and volitional skills*.

3   Theoretical Background

According to Weinert's notion of competency [9], competency models should describe relevant dimensions and skill components as well as a graduation of competencies which corresponds to different levels of learners' cognitive skills. The notion of competency encompasses *both cognitive and non-cognitive* skills and abilities. Motivational and volitional aspects and moreover social skills are subsumed under non-cognitive competencies. Competencies are of crucial importance for managing high demands and solving problems in complex situations.

So far the dimensions and components of the temporary competency model were derived from curricula and expert papers treating informatics modelling and system comprehension. At this juncture potential categories for modelling competencies were derived from a well-established software development process framework for instance. Thus, this step within the project was characterized by a theoretic and deductive methodical proceeding. However, a restriction on exclusively theoretically derived competencies would pose the danger that the described competencies do not sufficiently relate to *complex requirements in real situations.* Therefore, an additional step is necessary in order to determine competencies reliably, that is, ensuring an empirical access to the relevant competencies [6]. Conducting expert interviews represents an ideal empirical opportunity to detect the relevant competencies which belong to the two domains *informatics modelling* and *system comprehension*. Furthermore, a comprehensive qualitative analysis of these expert interviews guarantees further supplements, enhancements and rectifications of the competence dimensions if necessary.

4   Expert Interviews

4.1   Empirical Proceeding

Methodological Bases of the Expert Interviews. In each case two interviewers conducted the interviews with one interviewee. For economical reasons the interviews were conducted by telephone or Skype. For a successful interview no preparations by the experts were necessary. The interviews were standardized and their sequence is shown in the following:

(a) Welcoming and some small talk where it offered in order to create a suitable atmosphere and a foundation of trust.
(b) Mentioning the conditions of the interview, i. e.: Anonymity is assured, asking for permission for audio recording the interview in order to transcribe them in full, emphasising that the asked questions and given answers and descriptions are not understood as knowledge test but rather to explore an uncertain domain of competency.
(c) Referencing the underlying notion of competency (see above)
(d) Presenting the interviewing technique called *Critical Incident Technique* (see below).
(e) In each interview four hypothetical scenarios (see below) were presented and the interviewees were encouraged to give details of their personal proceedings and approaches solving the presented informatics problems.
(f) Quantitative evaluation of the four hypothetical scenarios by the experts with regard to representativeness of contents.
(g) Acknowledgement and farewell

The sample consists of a total of 30 experts on informatics, practitioners and theorists, with a division in three equal groups: experts in the domain of didactics of informatics, computer scientists and expert teachers (i. e. persons who are both teachers and teacher trainers). This composition of the sample was chosen in order to reveal the widest possible range of informatics and didactics expertise. The experts were addressed orally, by telephone or e-mail.

Critical Incident Technique. Having its roots in industrial and organizational psychology, CIT is used primarily to determine job requirements which are critical and decisive for acting successfully in occupational fields. The respondents are invited to delineate critical incidents belonging to a special domain. Thus, the critical incidents are gathered by conducting interviews in which the interviewees describe how they act in such situations and which conditions relevant to action are given. Thus, CIT strives for concise descriptions of those relevant skills, which are necessary to act successfully. Conversely, this means that diffuse and unsystematic circumscriptions which represent more or less common definitions are to be avoided.

Within the project MoKoM it was necessary to modify the mentioned procedure: The interviewees were not asked to remember critical incidents within their job, but they were presented with hypothetical scenarios containing such critical incidents. This methodically modified strategy was taken because the framework of the competency model has already been developed by means of theoretical considerations and on basis of plausibility. So, it was necessary to cover its competency dimensions and components by those hypothetical scenarios containing critical incidents concerning the domains of informatics modelling and system comprehension. A total of twelve hypothetical scenarios were developed. But in each interview only four scenarios could be covered. Of course, as varied as the scenarios are concerning the contents, the interview questions cover aspects of all four competency dimensions. An exemplary selection of hypothetical scenarios is presented in the following paragraph.

 - The scenario "Web-based game: Mensch aergere Dich nicht" targets system development. This game has to be implemented and consequently a detailed class diagram has to be designed.
 - Targetting system comprehension, the scenario "Car configurator" confronts with the task to test an unknown software.
 - The scenario "Exploration of a new standard software" is an example for system application.

Methodological Proceeding of Analysing the Expert-Interviews. The interviews were audio-recorded and transcribed in full. The interview transcripts were analysed by means of the qualitative content analysis according to Mayring [4]. The main objective of this method is to tackle enormous extents of text material. Three techniques can be distinguished: Qualitative content analysis can be used in order to *summarize* text material, to *explicate* text material and to *structure* text material. In the first case qualitative content analysis is mainly deployed with the purpose of compressing complex and extensive proportions of text material, i. e., the large text material shall be reduced to a manageable size in such a manner that important information is not wasted. In the second case additional text material is brought to those passages in the text which are necessary to interpret. In the third case it is attempted to extract a certain structure from the text material. This structure is brought to the text material in the form of a category system. All textual elements which are allocated by these categories are extracted from the text material systematically. The outlined main techniques of qualitative content analysis do not mutually exclude one another but rather complement each other well. Analysing the interview transcripts, all the three techniques have been applied. The theoretically derived competency model represents the mentioned category system which was applicated to the interview transcripts.

The standard procedure of qualitative content analysis proposes that the three techniques are applied as candidly as possible. Within the project MoKoM the deployment of the qualitative content analysis was characterized by an interpretative approach. That is, textual elements which are associated with knowledge, skills, abilities, motivation and social skills were extracted from the text material and assigned to matching competency dimensions and components of the competency model.

The results of qualitative content analysis are delineated in the following sections based on three interview transcripts. At first, each transcript was analysed separately by means of locating and capturing the textual elements which represent *meaning units*. Meaning units are such textual elements which contain relevant information concerning competencies, which means that they are decisive and critical for acting successfully in the outlined informatics domains. Moreover, further textual elements named *explications* were located which are associated with the gathered meaning units in order to explicate and depict them more concisely. In the next step it was investigated if the gathered meaning units and explications could be matched to the competency dimensions and components of the competency model. So, in a first step it is checked out which components of the competency model are addressed by the gathered meaning units and explications and which not. In addition to this it becomes apparent which gathered textual elements cannot be referred to the components of the competency model. In such cases the competency model has to be supplemented.

In the following step the three analyses were aggregated with the intent of obtaining an overall analysis.

4.2   Exemplary Illustration of Analysing the Expert Interviews

Within this chapter we describe how to refine our temporary competency model relying to the qualitative content analysis of the interview transcripts. In this context we will particularly focus on competencies concerning the development of informatics systems. Therefore we should first take a closer look at the respective competency dimension of our model.

Introduction Of The Competency Dimension. The competency dimension *basic competencies* comprises competencies in terms of *System Application*, *System Comprehension* and *System Development*. By *System Application* we understand competencies which enable learners to use an informatics system (IS) in a conscious and reflective manner instead of performing the trial and error method [5]. *System comprehension* covers competencies which empower learners to understand the internal structure of an IS and to rely it to the behaviour of an IS [8]. The competency component *System development* contains competencies which enable learners to develop an IS and to adapt it to new areas of application. A project oriented informatics education could be a feasible setting, within such competencies could become necessary e.g. when learners are asked to develop a small software project or to add new features to it. The subcomponents of system development are oriented towards the process workflows of the *Rational Unified Process (RUP)* [2] (amongst other approaches). These processes describe sequences of numerous activities as well as communication activities between relevant persons developing software. Additionally RUP illustrates appropriate models of Unified Modelling Language (UML) according to a specific workflow.

RUP sticks to principles of iterative Software Engineering and relies on component based architectures, which contain already existing components and new ones alike [2]. In order to be appropriate for the learners skills it (RUP) has to be tailored for didactic purposes. In this context System Development comprises *Business modelling workflow, Requirements workflow, Analysis & Design workflow* as well as *Implementation workflow* and *Test workflow*. The grading of this competency dimensions relates to the way of knowledge usage [1] (apply, comprehend and develop informatics systems) and the complexity of the informatics systems and tasks.

Illustration Of The Qualitative Content Analysis. By evaluating the mentioned interviews relying to the method of qualitative content analysis according to Mayring [4], we want to exemplarily demonstrate how to derive additional components from the interviews. Additionally we want to show first potential refinements of this competency dimension with regard to the results of the analysis. With respect to the focussed competency component *System Development*, a scenario comprising appropriate questions was developed in order to derive competency aspects in this specific domain [1]. In the following the scenario "'chat system"' including the respective questions will be introduced.

Scenario "chat system"
You are asked to develop an IT-based chat system. Within design workflow it is required to allocate program modules (e.g. java classes) to client and server.

Question 1: "What is your course of action to solve this problem? Which aspects do you have to bear in mind?"
Question 2: "Which kind of graphical models would you apply?"
Question 3: "Which cognitive skills are required to develop such a client server system?"
Question 4: "Could you imagine a potential pupil's procedure to solve this problem? What is the difference to your professional course of action?"[...]
At a subsequent stage of the software engineering process you are asked to establish a couple of small workgroups. In a next step you have to distribute tasks in order to empower the groups to contemporaneously implement the chat system.
Question 5: "Are there any attitudes or social communicative skills which are necessary for the group members to accomplish this problem? Could you imagine motivational aspects as well?"

---

[1] Additionally several different scenarios were developed for empirically deriving competency aspects concerning System Application and System Comprehension in particular.

Question 6: *"What do you have to bear in mind when you are planning to establish workgroups in informatics secondary education and you want to foster every pupil's modelling competence in equal measure?"*[...]

According to Mayring [4] we linked our temporary competency model to the interview material in order to derive the structure of the meaning units located in the interview transcripts. By posing *Question 1* with respect to the mentioned software development workflows we tried to investigate, whether the determined meaning units load respective competency aspects of this dimension.

In the following selected responses of the interviewees will be presented. These are consequently numbered according to the respective question. The first interviewee's response to *Question 1* is subdivided into *Answer 1a* to *Answer 1c*.

Answer 1a: "In the first instance I would think about what constitutes the system's functionality. In this connection I would have to consider which functional requirements are essential?

By performing a summarizing analysis [4] of the first interviewee's response we get *detection of functional requirements* as meaning unit. In the next step it has to be examined whether this meaning unit is interrelated to a competency component of the model. If this is not the case, this meaning unit might offer advice which aspects might be added to the competency model. Furthermore it indicates which competency components of the model have to be further differentiated. In this circumstance *detection of functional requirements* could be allocated to the competency component *requirements workflow*. This implies that this competency component is loaded by the respective meaning unit.

Answer 1b: "In a next step, I would consider how data transfer could be realized. Therefore I would ask myself the following questions: Which protocol could be appropriate? How to establish the connection between peers? How to communicate? What happens if the connection breaks down suddenly? What happens when the connection is closed normally by the user?"

To analyse the interviewee's utterances it was necessary to link a summarizing analysis to an explicating analysis. At first summarizing the whole passage of the interview transcript lead to *Design of communication and data transfer* as a meaning unit. In order to clarify, whether this unit is interrelated to the *design workflow* or to the *analysis workflow*, a closer context analysis had to be performed [4]. In this connection it was expedient to take the second sentence of this passage into account. At this juncture the interviewee explained that he/she wants to think about appropriate protocols to realize the chat system. Hence this course of action emphasises technical aspects and might finally lead to a design model "which serves as an abstraction of source code" [2, 11]. This kind of model intends to show, how the system will be realized in the implementation phase. Consequently the workflow presented by the interviewee can be linked to the competency component "Analysis & Design Workflow". Furthermore the derived meaning unit might eventually indicate the subdivision of *Analysis & Design Workflow* into *Analysis Workflow* and *Design Workflow*. If an advanced version of this competency dimension (after analysis of the whole amount of interview transcripts) should comprise these separated workflows, *Design of communication and data transfer* could be allocated to *Design Workflow*.

Answer 1c: "With respect to these preliminary thoughts, the respective function-alities can be allocated to client and server: The client implements GUI-classes, event management and the connection to the server; the server implements connection management, message processing and user administration."

By performing a summarizing content analysis we get *Client contains functionality to realise interaction with the user* and *Server contains connection-, message- and user-management* as meaning units. By "preliminary thoughts" the interviewee mentions his considerations concerning the Design workflow. Hence this meaning unit might be allocated to *Analysis & Design workflow* within the competency component. Furthermore the interviewee's assertion might show that learners have to develop the ability to select an appropriate course of action according to the respective iteration within the software engineering process and to choose serviceable modelling techniques. This assumption leads to *ability to choose adequate modelling techniques and course of action* as a meaning unit.

The *second question* is intended to gain additional information concerning informatics modelling and modelling techniques in particular during a software engineering process. More precisely we hope to get advice to expand the sub-components of "System Development" in terms of Mayrings structuring content analysis [4].

Answer 2: "As mentioned before, I would use UML class diagrams and UML deployment diagrams."

By summarizing this passage, we get *UML Class diagrams* and *UML deployment diagrams* as meaning units. As the interviewee did not distinguish between analysis class diagram and design class diagram the first meaning unit does not suggest any further differentiation of *Analysis & Design Workflow*. Although it can be allocated to this competency component.

The unit *UML deployment diagrams* can also be allocated to *Analysis & Design Workflow*. Furthermore the respective activity to *describe an informatics system by means of UML-diagrams* could be helpful to operationalise these competence categories.

The *third question* was posed in order to derive competency facets concerning informatics modelling in particular. More precisely it might help us to add competency components to the dimension *basic competencies* or to subdivide it as exemplarily illustrated before.

Answer 3: "As mentioned before, I would prefer an object oriented approach to solve this problem. Hence it is necessary that learners have acquired skills concerning object oriented analysis, object oriented design and object oriented programming as well. Furthermore they must be enabled to cope with a specific object oriented programming language. In addition learners have to notice, that they are faced with a distributed system. Hence they will have to acquire knowl-

edge concerning client server programming techniques. Marginally they have to be aware of event handling."

Summarizing the interviewee's utterances we obtain *oo-analysis, oo-design, oo-programming* and *knowledge of a specific oo-programming language* as meaning units. These can be assigned to the software engineering workflows of *basic competencies*. In this connection *oo-analysis* and *oo-design* can be allocated to *analysis & design workflow*. Analogously *oo-programming* and *knowledge of a specific oo-programming language* can be assigned to *implementation workflow*.

Later the interviewees were asked (*Question 4*) to describe their associations concerning pupils' course of action to solve this problem. By this means we want to investigate which competency components might be relevant for informatics secondary education. Furthermore we want to ascertain the distinction between pupils' and experts' proceeding to solve an ICT-related problem.

We intend to develop a competency model to categorize competencies concerning informatics modelling and system comprehension. According to Weinert's notion of the term competency [9], it comprises cognitive and non cognitive facets as well. Thus, in spite of deriving cognitive competency aspects from the previous questions, *Question 5* purposes to derive non cognitive competencies, i.e. attitudes, social communicative skills as well as motivational aspects. In order to perform a structural content analysis [4] it is convenient to take a closer look at the respective competency dimension of the model, i.e. *Non Cognitive Competencies*. It covers three competency components: *attitude towards an informatics system, social communicative skills* and *motivational aspects*. In the following it will be shown, how the interviewee's responses to *Question 5* might contribute to advance this competency dimension.

Answer 5: "In the first instance learners have to develop the willingness to bridge their lack of knowledge. Nevertheless they must have developed the attitude, that used technologies are purposeful. Another important attitude learners must have developed is, that they are working in the big picture instead of being faced with an isolated task with no interrelation to their environment. Concerning social communicative skills learners have to be able to reconsider concepts and ideas of other learners and to enhance them in creative manner."

In this case it is expedient to link a structural content analysis to a summarizing content analysis [4] in order to deduce potential meaning units and to allocate them to the respective competency components of *Non Cognitive Competencies*. In this context we get *willingness to bridge lack of knowledge, attitude, that used technologies are useful, attitude to work in the big picture, ability to reconsider other's ideas and concepts* and *ability to enhance other's ideas and concepts*. In this connection *willingness to bridge lack of knowledge* can surely be assigned to the competency component *motivational aspects*. The units *attitude, that used technologies are useful* and *attitude to work in the big picture* can be allocated to the component *attitude towards an informatics system*. After all the *ability to reconsider other's ideas and concepts* and the *ability to enhance other's ideas and concepts* can be doubtlessly understood as *social communicative skills*. Hence they can be linked to the respective competency component.

Question 6 purposes to investigate which non cognitive competencies are essential to solve such software engineering tasks in secondary education.

Preliminary Results Of The Qualitative Content Analysis. In summary the exemplarily demonstrated qualitative content analysis, which has to be understood as an intermediate result within a comprehensive analysis process, lead us to several meaning units. On the one hand meaning units referring to respective competency components showed us which components could be appropriate for a competency model (bold components within figure 1). These are *System Application*, *System Development* and the respective workflows, i.e. *Requirements*, *Analysis*, *Design* and *Implementation*. On the other hand they gave us a first impression how the competency dimension might be supplemented after finishing the whole analysis process, i.e. to add components to it (components within figure 1 marked by '*') or to further differentiate the components. With reference to this procedure it might be appropriate to separate *Analysis* and *Design* because there are technical meaning units which could be definitely allocated to *Design* instead of *Analysis* and vice versa. Nevertheless these worklows are interwoven and there are meaning units which could be allocated to both. Furthermore the analysis of a couple of interview transcripts has shown that it might be expidient to add the competence component *ability to choose adequate modelling techniques and course of action* (abbreviation: *Sequencing Pattern*). According to the respective software iteration, fostering this kind of competence should enable learners to choose a serviceable course of action and appropriate modelling techniques.

After the allocation of every obtained meaning unit the remaining (non addressed) competency components (grey components within figure 1) constitute components which have to be justified by taking further analyses of the interview transcripts into account. In this connection it has to be noticed that this exemplary scenario was intended to derive competency components concerning informatics modelling within software engineering processes. Another scenario emphasising the understanding of an existing informatics system might have lead to several meaning units which could be loaded to *System comprehension*. Beyound that they could be serviceable for operationalisation within the development of the competency measurement instruments.

5 Summary and Conclusions

The dimensions and components of the temporary competency model were derived by means of theoretical, rational and deductive considerations in a first step. For this reason the described empirical approach was indispensable in order to rectify and above all to substantiate and supplement them. The chosen proceeding consists of a special interview technique which deploys hypothetical scenarios presenting tasks on relevant competency requirements in a situated

Fig. 1. Left: Theoretically derived dimension — Right: First step towards an empirically refined and supplemented dimension

manner. The transcripts were analysed by means of qualitative content analysis, which means that the three main goals were realized particularly: summarizing, explication and above all structuring. On the basis of one exemplary hypothetical scenario, this article has shown how the interviews were conducted and analysed and how the results were applied to the competency model. On balance, this empirical approach is a useful complement of the first theoretical one. This means that a rectified, substantiated and supplemented competency model can be generated.

The described empirical proceeding is only one further step towards the development of science-based competency model because in spite of the advantages of the described methodological proceeding, restrictions are inherent in them. Thus further steps have to be undertaken to prove and validate the competency model. One methodological restriction is that relevant competency requirements can only be actuated fragmentarily. Thus, it is of decisive importance that the hypothetical scenarios contain representative, meaningful tasks and problems to solve. Subsequent to the scenario-based interview, the interviewees were asked to evaluate the scenarios with regard to its representativity and difficulty.

The behaviour patterns stated by the experts need not necessarily mirror their actual actions in a specific situation because they can be regarded as idealizations of perfect behaviour in a problem solving situation. In order to detect their actual behaviour and actions one would have to observe them while solving relevant informatics problems and to record their cognitive performances by loud thinking. For economical reasons it was abstained from that. On this issue, the different orientations of expertise of the interviewees serve as a corrective to some extent.

Of course, the deployment of the qualitative content analysis took place adhering comprehensible, methodological rules and principles. Nevertheless in some cases, during the procedure of content analysis it was inevitable to relinquish processes which were a kind of interpretative. In this way, depending on subjective decisions reduces quality criteria, that are, objectivity, reliability and validity. If the procedure of content analysis is characterized by explicit, concise and preferably detailed rules, principles and examples, the subjective scope of interpretation and sources of error will be reduced.

It is difficult to determine the quantitative levels of the required competencies by means of qualitative surveys and qualitative analysing. In order to take them into account it is necessary to deploy rating methods by means of which experts can evaluate in which grade competencies have to be present for a successful solving of relevant problems and demands. These problems and restrictions mentioned above raise the question of representativity of modelled competencies. That is, are the modelled competencies crucial, decisive and relevant for the target domains of informatics modelling and system comprehension? This remarkable issue will have to be adapted in further empirical procedures, for instance by using rating methods.

Concerning the further steps of research, the focus will be on content validity of the competency model. Informatics experts of different orientations should evaluate the relevance, difficulty, representivity and differentiation of the modelled competencies. Furthermore, the modelled competencies have to be proofed on criteria-validity. In this context, the focus of interest is following: Do the modelled competencies describe the requirements of successful acting in an actual and sufficient manner? In order to answer this question, instruments of measuring competency have to be developed with regard to the mentioned domains to operationalise the features of the competency model. Based on this, analysing the correlation between the levels of competency on the one hand and successful problem solving abilities on the other hand contains information on the criteria-validity of the competency model.

References

1. Anderson, L., Krathwohl, D.R.: A taxonomy for learning, teaching and assessing. A revision of Bloom's taxonomy of educational objectives. Longman (2005)
2. Rational Software Corporation IBM. Rational unified process. best practices for software development teams, white paper (1998)
3. Magenheim, J.: Towards a competence model for educational standards of informatics. In: WCCE 2005 - Proceedings of the 8th IFIP World Conference on Computers in Education (2005)
4. Mayring, P.: Qualitative Inhaltsanalyse. Beltz, Weinheim (2003)
5. Nelles, W., Rhode, T., Stechert, P.: Entwicklung eines kompetenzrahmenmodells - informatisches modellieren und systemverständnis. Informatik Spektrum (2009)
6. Schaper, N.: (Arbeits-) Psychologische Kompetenzforschung. In: Spoettl, G., Fischer, M. (eds.) Gegenstaende und Methoden der Berufsbildungsforschung, Berlin (2008)
7. Schlueter, K., Brinda, T.: Characteristics and dimensions of a competence model of theoretical computer science in secondary education. In: ITiCSE 2008, p. 367. ACM, New York (2008)
8. Stechert, P., Schubert, S.: A strategy to structure the learning process towards understanding of informatics systems. In: Working / Joint IFIP-Conference Informatics, Mathematics and ICT (IMICT 2007), Boston, USA (2007)
9. Weinert, F.E.: Concept of Competence: A Conceptual Clarification. In: Rychen, D., Salganik, L. (eds.) Defining and Selecting Key Competencies, Seattle (2001)

Having Fun with Computer Programming and Games: Teacher and Student Experiences

Clare McInerney

Lero - the Irish Software Engineering Research Centre, University of Limerick, Ireland
clare.mcinerney@lero.ie

Abstract. There has been a decline in the number of students studying mathematics, science and computing at third-level in Ireland. This may be because Ireland does not have a computing curriculum. However, in the UK higher education computing students fell in spite of the existence of such curriculum. According to teachers, student engagement and having a fun and exciting experience are essential ingredients when teaching computing to second-level students through a computing curriculum or other interventions. The Lero Education and Outreach program has developed a set of materials "Having fun with Computer Programming and Games" for second-level students aged 15-16. The goal is to engage students in computing and to offer them an enjoyable and positive experience. Working with two second-level ICT teachers, we designed a set of materials during summer 2008 that have been deployed in a number of schools.

Keywords: Second-level outreach, second-level curriculum, computational concepts, computational thinking, engagement.

1 Introduction

Students in Ireland choosing mathematics, science and computing[1] declined by 13 per cent in 2008.[2] Despite emerging opportunities and a growing need for qualified IT people in Ireland and globally, students are not signing up for undergraduate programmes in computer science [1], [2], [3].

The report [4] published by Forfás, Ireland's National Policy Advisory Body for Enterprise and Science, highlights the need "to reverse the recent decline in the domestic supply of high-level computing" graduates.

The recent report commissioned by both the Higher Education Authority of Ireland and Discover Science and Engineering [5], states that encouraging more young people to take up a career in computing and information technology cannot be addressed in a single communications campaign, but that it needs to be tackled in a variety of ways. In the same report, it is suggested that students have a limited exposure to information technology in schools and find it difficult to find a role for information technology in the educational environment.

---

[1] Computing refers to software development in the wider sense.
[2] http://www.siliconrepublic.com/news/article/13571

J. Hromkovič, R. Královič, and J. Vahrenhold (Eds.): ISSEP 2010, LNCS 5941, pp. 136–142, 2010.
© Springer-Verlag Berlin Heidelberg 2010

The following sections will describe:

- Pilot rollout of the materials developed
- Initial reactions from teacher and pupils
- Plans for the future

2 Materials

"Having Fun with Computer Programming and Games" teaches computational concepts using Scratch [6] and computational thinking [7] to students. We also use Computer Science Unplugged [8] materials to teach searching methods to students without using computers. We have designed 45 hours of teaching material that teach students how to build computer games, create animations that use art and music, design, work in teams, test components of their games, modify and add new functionality to existing programs, present their projects to their classmates and provide feedback to their peers. The materials developed are available at http://www.lero.ie/educationoutreach/secondlevel/scratchlessonplans under a Creative Commons license.

The materials consist of 10 modules with each module containing 2-5 lessons. The lessons have been designed by experienced teachers for teachers that do not normally have a computer science background. Following the Irish National Council for Curriculum and Assessment (NCCA) guidelines for such courses, the materials teach the following key skills:

Information processing

- Researching and recording information on the Internet for research projects requires students to process large amounts of data in order to evaluate and extract appropriate information.

Critical and creative thinking

- Providing feedback to classmates by critically evaluating student research projects. They have the opportunity to incorporate creative thinking into their Scratch projects and research projects.

Communicating

- Presenting a research project and a Scratch project to the class; working in teams to build a Scratch project. Providing feedback and suggestions to classmates.

Working with others

- Planning and designing a Scratch project gives students an excellent insight into working with others, organising and delegating work within a group, taking responsibility for tasks and completing them.

Being personally effective

- Writing computer programs, animations, games and stories gives students a great sense of achievement as they are able to incorporate their ideas and designs and implement them into a finished product.

3 Pilot Project

The Lero Education and Outreach Programme is targeting the optional year between the second-level junior and senior cycle: transition year. Transition year is unique in that it is an exam-free, stress-free year and it "encourages the development of a wide range of transferable critical thinking and creative problem-solving skills."[3]. The Irish secondary school system is described in more detail in [9]. The "Having Fun with Computer Programming and Games" materials were rolled out in fifteen schools in geographically distributed areas of Ireland. We conducted interviews with five teachers that taught the materials during the 2008-2009 school year.

Table 1 shows the numbers of boys and girls that enrolled and completed "Having Fun with Computer Programming and Games". It shows the total number in the group, the number of weeks and the number of hours assigned to teaching the materials. It is clear from the table that timetables and scheduling differ significantly across schools in Ireland. A Recommended Timing for Modules [10] document is available to guide teachers.

Table 1.

	Number of Boys	Number of Girls	Total In Group	Duration (Weeks)	Duration (Hours)
School 1					
	0	28	28	32	67
School 2					
Group 1	0	23	23	8	10
Group 2	0	22	22	8	10
Group 3	0	25	25	8	10
Group 4	0	20	20	8	10
School 3					
Group 1	12	6	18	16	29
Group 2	7	8	15	16	29
School 4					
Group 1	10	10	20	8	25
Group 2	12	11	23	8	25
Group 3	9	11	20	5	17
Group 4	13	9	22	6	20
Group 5	13	9	22	6	20
School 5					
		13	13	32	42
Totals	76	195	271		

In school 3 and school 5 "Having Fun with Computer Programming and Games" is an option.

---

[3] Department of Education Transition Year Support Service http://ty.slss.ie/

3.1 Computational Concepts Using Scratch

When designing "Having Fun with Computer Programming and Games" materials, we evaluated a number of software tools that could be used to teach computational concepts to second-level students. For instance, we considered Greenfoot [11]. It is an excellent tool for teaching programming and freely available; but given the limitations of technology infrastructure in Irish secondary schools (a JVM generally requires more than 256MB of RAM) and lack of Java programmers amongst second-level computer teachers, the tool was deemed unsuitable for the Irish secondary school system.

Scratch is being used successfully to teach programming skills to novices. It has been used as an introductory language at undergraduate level [12, 13]. We are using it at second-level schools in Ireland. Scratch is freely available and easy to install and use. Scratch teaches computational concepts to students in a fun and engaging way. Teachers reported that student engagement with Scratch was "far superior"[4] to the level of engagement while studying ICT literacy skills and that "motivation levels were high among all students".

In Module 1 we show students how music can be incorporated into Scratch. Students take part in a battle of the bands competition. This task was very engaging for students. In school 4 the teacher remarked that "this was very popular and we had a big band jam on the projector with the speakers turned up".

In Module 2 we introduce algorithms. We asked students to design an algorithm on paper to draw a picture. Students enjoyed this task. Students were introduced to turtle graphics and drew squares, circles, etc... However, once students started working with more complex shapes that used nested loops, "students appeared unimpressed". This lesson lost the attention of students and teachers were obliged to continue with other materials.

In Module 5 Scratch cards are distributed to students and they are required to implement solutions to various problems. There are three levels of Scratch cards: easy, difficult and extreme. Solution cards are provided for teachers. The cards are used to revise all computational concepts that have been taught. Easy cards were completed by all students and difficult cards were completed by many students. While the teacher in school 1 remarked that the extreme cards were "too difficult to do alone", the teacher in school 4 commented that the "level of tasks and their challenge was very appropriate" at the extreme level.

In Module 6 students are given a code maintenance task. They are presented with existing Scratch projects and asked to make modifications. Teachers reported that this task "gave them confidence". The teacher in school 2 commented that "Module 6 was the most successful module with students. It allowed less able students to complete minor modifications and more able students to complete more advanced modifications". The teacher in school 5 indicated that "more modifying code lessons would be welcome."

In Module 10 students design and implement a final project in groups. They are required to fill out a project worksheet to document their work as the project progresses. They present their project to their classmates when the project is completed. In terms

---

[4] All quotations in "" are taken verbatim from teachers in one of the five schools studied.

of presenting the completed product teachers reported that "the groups were very enthusiastic about displaying their work". The teacher in school 3 reported that "some students showed good initiative and worked on parts of their project at home".

Irish secondary schools have open days for primary school students. In school 3 a number of students presented Scratch and their projects to prospective students and according to the teacher involved "this was very well received both by parents and prospective students".

Based on teacher feedback, students and teachers had a positive and engaging computing experience learning computational concepts in Scratch.

3.2 Computational Thinking

The "Having Fun with Computer Programming and Games" materials include computational thinking. In her computational thinking paper [7], Jeanette Wing suggests that when problem solving we might ask "How difficult it is to solve?" In our materials we present students with complex problems and solution implementations. We want students to gain an understanding about the complexity of problem solving. We are not asking students to build solutions to the complex problems, but manually solve a complex problem game. By presenting complex problems in this way, we want to give students an appreciation for the grand challenges in computer science.

In Module 7 lesson 1 we present students with an implementation of the Towers of Hanoi[5]. We ask them to solve the problem for 3 disks. The teacher explains how the number of moves required to solve the problem increases significantly as the number of disks increases. The teacher in school 3 reported that "there was great competition in class to see who could solve with 3 disks. A few students were able to solve the problem in 7 moves and moved onto 4 disks." In Module 7 lesson 2 we present students with an implementation of the Traveling Salesman Problem[6]. Students are asked to find the optimal tour for 10 cities. Again we explain how the complexity of the problem increases as the number of cities increase. The teacher in school 4 noted "the Traveling Salesman Problem was less popular than the Towers of Hanoi game. Only the more persistent, mathematically-minded pupils found the activity engaging".

Feedback from teachers indicates that it is possible to interest and engage a wide variety of students in understanding complex computer science problems by allowing them to solve a complex problem game. The competitive aspect of games is appealing to students. Teachers reported that students enjoyed the computational thinking module.

3.3 CS Unplugged

The "Having Fun with Computer Programming and Games" materials include a lesson from Computer Science (CS) Unplugged. CS Unplugged, based at the University of Canterbury in Christchurch New Zealand, is a well established project running for over 15 years. CS Unplugged teaches computer science without a computer. CS Unplugged teaches "principles of computer science such as binary numbers, algorithms and data compression through games and puzzles that use cards, strings, crayons, and lots of running around"[8]. In Module 3, we use CS Unplugged materials to teach

---

[5] http://nlvm.usu.edu/en/nav/frames_asid_118_g_3_t_2.html
[6] http://www.tsp.gatech.edu/games/tspOnePlayer.html

searching algorithms. Linear search, binary search and searching using hash tables are taught in a battleships game. The teacher in school 3 commented that "students really enjoyed the Battleships exercise. There were 1 or 2 moments of confusion with some students but nothing significant and the exercise proved to be good fun".

Feedback from teachers reveals that students interacted well in their teams during the Battleship exercises and completed the searching algorithm games successfully.

4 Conclusion and Future Work

According to teachers interviewed this first time experience of computing was engaging and fun for students. "Having Fun with Computer Programming and Games" offers a positive and exciting experience to computing novices. To address the issue of students loosing attention during particular lessons, we made modifications to the materials during the summer of 2009. For example, we added more detailed steps and guidance in the nested loops lesson and we added a two player Traveling Salesman Problem game in the computational thinking module to include a competitive element.

Because the materials were designed for teachers that do not normally have a computer science background, extensive training is not required for teachers wishing to teach the modules to transition year students. This allows us to roll out the materials nationally to all schools in Ireland in September 2009. Lero has formed partnerships with CIO Ireland, a group of senior ICT executives, and the Institute of Technology Tallaght (IT Tallaght) and the Irish Computer Society (ICS) to support the national rollout.

In terms of increasing the uptake of computer science undergraduate courses, transition year students that participated in the pilot project will be making their university choices in August 2011. At that point we will know if "Having Fun with Computer Programming and Games" has had an impact on student choice of course at third-level. In the meantime we plan to survey students that complete the course. We hope that the uptake of the materials will increase nationally as we move from the pilot phase to a national rollout and that new students and teachers will have a similar positive experience to students and teachers involved in the pilot project of 2008-2009.

Acknowledgments. This work was supported, in part, by Science Foundation Ireland grant 03/CE2/I303_1 to Lero—the Irish Software Engineering Research Centre (www.lero.ie). The author would like to thank Prof. Kevin Ryan and Dr. Jack Downey at Lero for helpful comments and suggestions on this paper. The author is grateful to the five teachers interviewed.

References

1. A Study on the IT Labour Market in the UK, A Report by Research Insight commissioned by CPHC (June 2008), http://www.cphc.ac.uk/docs/reports/ cphc-itlabourmarket.pdf (accessed August 2008)
2. Vesgo, J.: Continued Drop in CS Bachelor's Degree Production and Enrollments as the Number of New Majors Stabilizes. Computing Research News 19(2) (2007)

3. Teague, J.: Personality type, career preference and implications for computer science recruitment and teaching. In: Proceedings of the 3rd Australasian conference on Computer science education. ACM, The University of Queensland, Australia (1998)
4. Forfás: Expert Group on Future Skills Needs, Future Requirement for High-level ICT Skills in the ICT Sector, http://www.forfas.ie/media/egfsn080623_future_ict_skills.pdf (accessed August 2008)
5. Ipsos MORI: Career Opportunities in Computing & Technology in Ireland, http://www.hea.ie/en/webfm_send/2241 (accessed August 2009)
6. MIT Media Lab, Scratch, http://scratch.mit.edu (accessed October 2009)
7. Wing, J.: Computational Thinking. Communications of the ACM 49 (2006)
8. Computer Science Unplugged, http://csunplugged.org/ (accessed October 2009)
9. McInerney, C., Hinchey, M., McQuade, E.: Investment in Information and Communication Technologies in the Irish Education Sector. Education and Technology for a Better World, 83–91 (2009)
10. Recommended Timings for Modules, http://www.lero.ie/download.aspx?f=Recommended+Timings+for+using+Modules.pdf (accessed October 2009)
11. Greenfoot, http://www.greenfoot.org/ (accessed October 2009)
12. Malan, D.J., Leitner, H.H.: Scratch for budding computer scientists. In: Proceedings of the 38th SIGCSE technical symposium on Computer science education. ACM, Covington (2007)
13. Wolz, U., Maloney, J., Pulimood, S.M.: 'scratch' your way to introductory cs. In: Proceedings of the 39th SIGCSE technical symposium on Computer science education. ACM, Portland (2008)

Showing Core-Concepts of Informatics to Kids and Their Teachers

Roland T. Mittermeir, Ernestine Bischof, and Karin Hodnigg

Institut für Informatik-Systeme
Alpen-Adria Universität Klagenfurt
Österreich
{roland,ernestine,karin}@isys.uni-klu.ac.at

Abstract. Computer science education in schools focuses in many countries on the use of IT-equipment rather than on foundational concepts informatics offers. Although this is partly justified by immediate use of IT-literacy of pupils, it is as much due to the lack of deeper knowledge in computing on the part of teachers. But this application-based instruction distorts the pupil's perception of informatics and thus, leads to lack of interest in informatics related professions.

This paper reports on a project that shows both, pupils and teachers, some principles of informatics in order to make the students curious to learn more about it and showing both groups that informatics offers more than its applications. It also shows teachers that concepts of informatics are not too difficult to teach and to grasp.

1 Motivation

With the proliferation of PCs, informatics instruction in Austrian schools (as well as in many other countries) degenerated to courses on how to use PCs. This leads to a distorted image of the discipline in the minds of the young generation. There are several causes for this effect, and again, while speaking about the situation in Austria, we are convinced that the situation might, with gradual differences, also characterize the situation in other countries:

— Computer Science / Informatics[1] is a dynamic discipline. To cope with its dynamics, curricula are formulated on a rather general level, leaving room for creativity and interpretation by teachers.
— The majority of school-teachers currently teaching informatics courses have obtained the academic education in other disciplines than informatics. Thus, their knowledge in informatics has been acquired by means of in-service courses which are rather application oriented. Hence, these colleagues do not possess a profound knowledge in core topics of the scientific discipline of informatics.
— Therefore, it is easier for them to teach shallow, application-related concepts than deeper principles offered by Computer Science.

---

[1] In the sequel, we treat the terms *Computer Science*, *Computing*, and *Informatics* as equivalent.

J. Hromkovič, R. Královič, and J. Vahrenhold (Eds.): ISSEP 2010, LNCS 5941, pp. 143–154, 2010.
© Springer-Verlag Berlin Heidelberg 2010

— However, it is relatively difficult to re-educate those in-service teachers, making them competent at such a profound level that they can show pupils the beauty and the intellectual reward of informatics.
— Given this situation and given the curricular requirement that informatics instruction has to lead to computer literacy, far too many teachers orient themselves on the ECDL curriculum in a rather strict manner. This gives pupils the chance to obtain an external certificate, but it does not motivate teachers to delve into those parts, where informatics education will become interesting even in school.

The situation described above lead us to devise an agenda for bringing "true informatics concepts" into school aiming at both, students and their teachers. This paper first describes the ideas behind this approach and the overall didactic design of the project *Informatik erLeben* (roughly translated: *experiencing informatics*). Then we describe the intended setting of our interventions in classes. The ensuing interim evaluation is based on repetitive interventions in some partner schools.

2   Guiding Ideas

2.1   Shifting Educational Aims

Informatics education has been introduced into the general curriculum of Austrian schools in 1984 [12]. Focus of instruction was programming. Those who ventured on educating themselves in informatics and passing this education on to their pupils were excited about the new technology and the intellectual challenges it offered. They focussed on programming and thus passed algorithmic concepts to pupils. Possibly they overestimated the importance of programming skills for the general public and possibly they underestimated the difficulty it poses for some pupils. The excitement of these innovators might have made them blind towards justifying why and what should be taught. Hence, the overall educational value of informatics education was rarely discussed.

As technology evolved, it became more powerful and cheaper. Most households with children of school-age now have a PC at home. It is equipped with pre-installed office software (notably for text processing) and it provides access to the internet, serving as communication infrastructure and as information resource.

To make use of these resources, people had to become computer literate. *Computer literate* is an interesting term, since it raises mastering of certain computer-related skills to the level of an important cultural skill, comparable to the capability to read, (write), and to comprehend. But in contrast to general literacy, where no accepted yardstick exists, the ECDL Foundation provided with its European/International Computer Driving License (ECDL/ICDL) an educational standard for computer literacy [5]. Both, the international scope of the certificate, as well as the fact that it is obtained on the basis of an examination before a school-independent board, made it interesting for industry and consequently also for parents. Schools offered their students auxiliary courses to prepare themselves for the ECDL-exam[2]. However, this

---

[2] Some schools, specifically technical ones, prepare their pupils also for other external exams such as CISCO or Microsoft Certified Engineer.

seemingly positive medal had also a backside: the success of the ECDL in Austrian schools caused informatics instruction to degenerate to an instruction in computer use. But teaching IT-use under the term of *Informatics* leads to a distorted image of informatics as scientific discipline. This distortion can be observed not only with pupils but also with a substantial number of teachers. Showing the intellectual merits of mastering certain aspects offered by informatics [4, 13] requires mastering some deep concepts of the discipline. In contrast, to teach how to do text-processing, how to prepare slides or to send e-mails requires just knowing some commands of the respecttive software.

Consequently, this curricular shift was welcomed by many teachers. Arguments against programming were put forward, claiming that it is difficult and it cannot be the duty of school to make all pupils programmers. While the subject is still called *Informatics*, its content is *Applications of Information Technology*. Micheuz, asking in a recent survey [10] ninth-graders (the age bracket with compulsory informatics education in Austria) about terms characterizing their informatics instruction, obtained as dominant terms *Word, text processing, Excel, boring, presentations*. Terms such as *enthralling* or *interesting* appeared only after a marked distance in the frequency count. This observation, in conjunction with the result of interviews conducted at two schools [1], lead us to devise a programme showing pupils the difference between using products stemming from informatics based developments and the scientific discipline of informatics [3].

2.2 Where to Start from?

Being university teachers, it was obvious that we had to address the interface between school and university studies, i.e., the upper grades of secondary education. But an advisory board told us that this might be too late. We better start at a level where child-like curiosity is still dominant and where interests are not yet firmly defined (or spoiled). Thus, we had the challenge to introduce informatics in primary education.

While some approaches to make informatics concepts comprehensible for children without using computers do exist (e.g.: [2, 7]), we felt that it would be insufficient to just copy them. We were striving for a set of interventions that could be pursued throughout the scope of school-education and that could be adopted by teachers of the respective age groups.

This set contains nine different concepts of informatics stratified in such a way that units are suitable for classes in the age group of 8 up to 18 years. For each unit about two hours of instruction are foreseen.[3]

But as pupils change classes and pupils change schools, we cannot hope to accompany a given class from 3rd level primary till 8th level secondary school (12th grade). However, we might accompany some of them, and moreover, we do accompany their teachers. What might be a singular intervention for a particular pupil will develop in a sequence of interventions for teachers and if the teachers like what they have observed they might try it on their own. This facet of these interventions is important as it perceives the teacher as a natural multiplier for our idea of active and interactive units. To encourage them, the booklet [3] announcing the approach is rather thin. The body of the material can be downloaded from a web-site[4] which is still evolving.

---

[3] In the meantime units for pre-schoolers were developed. These last only 60 minutes or less.
[4] http://informatik-erleben.uni-klu.ac.at/

On purpose, these units are not necessarily focussing exclusively on informatics concepts. They rather funnel towards informatics. Especially for the young age group, we focus on aspects of science and technology, e.g., combining them with art to arrive at computer graphics and colour prints.

Physical action plays a key role in this concept. *"Unplugged"* is another key term. Every unit is designed in a way to provide pupils with a motivation or a given problem, some (often very little) theoretical input, and a simulation and/or animation of the underlying technical or informatics-based concept. The pupils have to get up and actively take part in that simulation (for small children of game-like nature) or actively observe the activity other pupils are performing. E.g., as preparation for programming, we ask them to write an essay for giving directions [9] in order to train the preciseness of their verbal expression independently of any programming language syntax and then execute the prescriptions written in those essays. Further on, all units strive that doing and observing what is or has been done is kept in balance. We consider this to be an important principle, since most inventions and innovations are not the result of pure introspection but rather the result of careful observations and the capability to relate seemingly unrelated observations.

This important principle is highlighted by the term *"Informatik erLeben"*. In German, something *zu erleben* means *to experience* something. *Leben* means *life*. By asking pupils to play the role of components of computers or being manipulated as data elements by some algorithm, we attempt to counter the argument that informatics deals just with abstract concepts and is therefore difficult to teach. Thus, not only the abstraction hurdle is overcome, the execution of algorithms is also slowed down to human scale. Further pupils and teachers should see that informatics is not just something that has to do with an information technology-based machine. Informatics has to do with reflecting upon and structuring information related artefacts and concepts. In general, this has to do much more with humans than with the machine that processes images of these artefacts and concepts encoded in bytes and numbers. Bringing the notion of school-informatics slightly away from "dealing with computers" towards "designing applications for humans" is an important target also in terms of correcting gender-specific misconceptions reported in [1].

A further assumption behind the project is that young people are naturally curious and naturally interested in demonstrating their strength at least in those areas they are interested in and they like to identify themselves with. Sports activities or music may serve as witness against too pessimistic counterarguments.

When distributing the booklet [3] to teachers, quite a number of administrators warned us that it contains nice, but very difficult ideas. However, we were hoping that presenting informatics concepts in a manner suitable for the respective age group an initial impulse will be given to teachers, showing them both:

a. Informatics is not all that difficult. It is feasible to engage their pupils even in intellectually challenging tasks, if these tasks are presented in such a way that the kids (or youngsters) can follow and enjoy them.

b. The kids (youngsters) are not basically technology-averse. If these technical aspects are well presented, they will enjoy them and even use some of the concepts for private games.

3 Description of Some Units

Currently, the project *Informatik erLeben* comprises a range of nine basic informatics topics. Each topic consists of several units and modules which can be combined with modules from other topics (e.g. the game from the Morse-code can be combined with coding as well as with network topics). Whenever appropriate, ideas from [2] or [7] guided us or were integrated into modules. In the sequel, related strings of topics are described in an exemplary manner.

3.1 Image Processing and Graphics

Even though it is not a core topic of informatics, image processing and graphics are subjects that can be easily discussed with very young children. The modules do not only communicate how images are represented in the computer but also try to show which physical principles are important for yielding and in mixing colors. Children in primary school classes can mix subtractive colors as done by printers. Physics questions, like *what is light, what does light consist of* are discussed. On this basis one can explain how pictures can be represented in computers.

Raster graphics are introduced by a communication game. The pupils get a handout with a raster and paint a simple image by filling some boxes. This image is mapped into a 0-1-pattern. The resulting sequence of numbers is sent to colleagues who are to reconstruct the image. This sets the ground for introducing vector graphics. Color depth is discussed with pupils from lower secondary school upwards.

3.2 Coding

Coding, the assignment of meaning to a signal, letter, or number, is a fundamental concept of informatics. With the coding lessons, this concept can be taught in a playful way by the Morse-game [7]. This game is based on the idea of identifying a partner via light signals emitted by flashlights. We use the Morse-code as primary example, but many students are already aware of sign language or Braille characters. After kids have invented their specific classroom code expressing each letter of the alphabet by a sequence of colored spots and writing short messages like their name in this code, even 3rd graders easily grasp the fundamental idea of coding, the mapping between letters of different alphabets. With explanations adequate for the age of children, even code trees can be used and are well-understood. In secondary school, advanced coding aspects like Huffman codes or code optimization are discussed. Error detection and code correction are explained using games.

3.3 Encryption

Simple encryption algorithms such as Caesar cyphers can be taught already to primary school pupils. The idea of having a key and hiding messages from each other is a very important aspect in modern society. Beginning from the idea of secret codes, a simple Caesar chiffre is used to show the concept. Already primary school pupils can handle and use this encryption algorithm. In secondary schools, students are already capable of handling more advanced and elaborate algorithms like RSA or public key encryption.

3.4 Hardware

Whenever one thinks of computer science, the underlying hardware comes to mind. Although computers are used a lot in schools (laptop classes, computer supported classes, ICT in school), hardly any pupil knows how a computer works internally or what it consists of. Thus, a spectrum of age-specific units was designed to provide insight into how the black-box computer really works. Progressing from the execution of a simple addition task for primary school to more complex problems for secondary school, the functionality of a computer is animated and illustrated. In these units, pupils assume the role of different functional units. Serving as memory, registers, command counter, ALU, etc., they cooperatively execute the tasks assigned to these units when running a given program. To deepen the insight, the students are afterwards encouraged to open, dis- and reassemble a PC. All components are named and explained. Pupils are very attentive during this task and enjoy the sensual experience of touching hardware components.

3.5 Algorithms, Searching and Sorting

Searching and sorting are concepts suitable to instill algorithmic thinking. Hence, units animating these concepts and evaluating different strategies were developed alongside units leading towards programming.

The most elementary unit in this sequence starts with pre-schoolers who are to blindly select the longest (shortest) pencil in a bag. The issue here is that they have to "invent" an algorithm to solve the problem and verbalize it afterwards. With pupils of higher age, precision of expression and the notion of algorithmic concepts are introduced by having them give directions to a visitor and then blindly follow the path so described within the school building. Descriptions are firstly given in natural language and then in structured natural language.

With searching and sorting, one can easily motivate the need for efficient solutions, e.g., by introducing some emergency situation and then contrast two different searching or sorting algorithms. The algorithms are animated and played with some students serving as data to be manipulated. Other students are acting as algorithmic guards executing the given algorithm on the "data" or as observers counting the number of operations. Initially, students are placed in a row by size, like in gym classes. Sorting (or search) criteria might be birthdates of the students or their last names. The respective algorithms are executed on the students and contrasted (e.g.: Bubblesort or Selectionsort with Mergesort (for younger kids) or Quicksort (for upper secondary school)). This animation of an algorithm shows first of all the idea behind a given algorithm and contrasting two algorithms introduces the notion of algorithmic complexity as well as the concept that in certain situations, obtaining a correct solution is insufficient. One needs to have the result in time. Combining searching and sorting shows the effect of a data structure on the complexity of the solution to a problem. Establishing a search tree links searching with recursion.

3.6 Operating Systems and Computer Networks

The different role of the operating system and programs running under its control are highlighted for younger pupils by analogies to a pharmacist serving customers.

Depending on knowlede about hardware units, scheduling strategies and eventually problems of deadlock or lifelock are discussed.

With networks, concepts of politeness introduce the basic principle behind protocols. On this basis, net topologies and specific protocols and protocol levels are discussed.

4 Intermediate Evaluation

The project started in winter 2008. Spring served as testing period and proof of concept with four schools. Currently, there exists a partnership with 11 schools comprising 17 classes. Each class obtains 3 interventions and is also invited to visit once an IT-focused company and once the CS-department of the university in order to see IT-work-places in industry and science. This partnership lasts for 3 semesters. But as 7 of these classes (4th grade primary, 9th grade gymnasium) were to be dissolved or restructured in summer '09, they run the project on an accelerated schedule and completed it already in July '09.

4.1 What and How to Evaluate

Since our main aim was to change attitudes with both pupils and teachers, a typical quiz with knowledge- or skill-related questions would be inappropriate for evaluation. A true assessment could be done only after observing that some teachers adopted concepts observed by us and that students opted for certain IT-related studies or professions. Observing this effect will take years. Therefore, other strategies had to be resorted to. The following evaluation strategies were considered:

- questionnaire with two questions to be answered in free text,
- creation of questions to the topic by the children,
- observation of the pupil's attention,
- review session with teachers and pupils.

4.2 Free-Text Evaluation on Grasping the Concept of Informatics

The evaluation approach using free-form questions aimed straight towards the question whether our interventions changed some concepts associated with the word *Informatics*. A collection of questions had been defined a priori. Out of this set, two selected questions were asked before the first and after the last *Informatik erLeben* unit (a time span of about two or three months, depending on the class).

The collection of the questions consisted of:

- What can you do with informatics?
- Who in your class is best in informatics? Describe his/her qualities.
- What do you consider to be the most interesting aspect of informatics?
- Where are you successful with informatics?
- How do you imagine the workplace of a computer scientist?
- If you were a computer scientist, what would you do on a normal workday?
- Informatics is for me to…

It appeared that not all of the questions were suitable for this aim. The question about the workplace for example was interesting but didn't show a significant change of the attitude towards informatics. Other questions, like *What does informatics mean for you?* showed for some of the pupils (in some classes about half of them) that informatics was after the unit more than using the computer. Before our intervention informatics was only using the computer and its programs, using internet, communicating via internet and playing games on the computer for all pupils in all classes (even in upper secondary schools). A change in perception is also evident by answers to the question *What can you do with informatics?*

Table 1. Frequency of answers before (left) and after (right) 3 interventions to the question "What can you do with informatics?"

Playing computer games	20	Decompose computers	12
Surfing the internet	17	Games	9
Downloading films	11	Codes/coding	7
E-Mail	10	Morse-alphabet	7
Watching and processing pictures	10	Internet	6

Table 1 reports on the left side the answers of 23 primary school pupils before having attended a unit. After three units of *Informatik erLeben,* the same group tried to include some of the concepts they have been confronted with during the lessons but we still find terms like "playing" or "internet" in the answers.

4.3 Creating Questions as Evaluation Approach

Based on the assumption that the quality of questions correlates with the asker's knowledge, we wanted the children to form questions to the topic before and after the *Informatik erLeben* units. This approach was tried in primary schools as well as in lower secondary schools. But in both cases it was not successful. Apparently, the universe for asking questions related to informatics is too broad and moreover, the culture of noting and asking questions is not yet established in such a young age group. Before the unit some pupils had a few questions but there was no connection to the topic, possibly because they knew too little about it. After the unit, almost no new questions occurred.

4.4 Attention as an Indicator of Attractiveness

This approach [8] involves teachers observing their students and thus caters for the aim to address both groups by our didactical interventions. Based on the assumption that pupils are attentive only if they appreciate the situation, the results of this observation indicate whether a pupil liked the lesson or not. To assess the pupils' attention teacher were asks to observe their pupils and to note in a questionnaire each pupil's attention according to the following scale:

— On-task passive: pupil follows the lesson passively,
— On-task active: pupil follows the lesson actively on his/her own,

− On-task reactive: pupil follows the lesson actively by reacting to a question,
− Off-task passive: pupil doesn't follow the lesson, but doesn't disturb,
− Off-task disturbing: pupil doesn't participate to the lesson and disturbs.

The questionnaire was divided into four categories according to the different approaches occurring in most units (individual work, pair/group work, lecture, and animation). With multiple entries, teachers could express that a pupil showed a spectrum of attentiveness throughout some activity. In completing the form, teachers have to be attentive on what is going on in class and form their own opinion as to whether the approach followed by the intervening person is attractive and suitable for the class.

The evaluation the observation forms showed that most pupils were on-task. But as shown by the ensuing tables, age-dependent differences could be noticed.

Because the units presented to different classes varied, not every module or unit contains all four teaching approaches. Therefore, the total number of pupils observed on particular task categories varies (the respective total is indicated in the column heading of the tables below).

Table 2. Activity report for 4th graders

	Lecture (n=64)	Observation and animation (n=62)	Individual work (n=51)	Pair/group work (n=62)
On-task passive	1	7	0	0
On-task active	58	52	51	61
On-task reactive	25	0	22	24
Off-task passive	0	3	0	0
Off-task disturbing	2	0	0	1

As shown by Table 2 most children from primary school classes were on-task active or reactive, in all four categories (methods). One can say that pupils of primary schools were in general open to any of the topics presented. They were very attentive and actively participated in the units.

Table 3. Activity report for students of lower secondary schools

	Lecture (n=54)	Observation and animation (n=50)	Individual work (n=48)	Pair/group work (n=13)
On-task passive	24	22	10	0
On-task active	30	26	38	13
On-task reactive	0	1	0	0
Off-task passive	0	1	0	0
Off-task disturbing	0	0	0	0

Pupils from the lower secondary schools were attentive too, but didn't participate as actively as the younger pupils (see Table 3). While the majority of them was on-task active, a substantial portion was observed to be passive. Apparently, the interventions met already with stronger preconceptions concerning fields of interest of the young person.

Table 4. Activity report for students of higher secondary schools

	Lecture (n=26)	Observation and animation (n=15)	Individual work (n=6)	Pair/group work (n=18)
On-task passive	11	8	2	5
On-task active	6	5	6	13
On-task reactive	3	2	0	0
Off-task passive	4	0	0	0
Off-task disturbing	2	0	0	0

Till now we worked only with three groups of the higher secondary schools (9th graders). The results reported in Table 4 are similar to those witnessed with lower secondary schools. However, the trend persists that interest for a subject or lack thereof is broader stratified in higher age group.

4.5 Review Session with Teachers and Pupils

For the classes terminating the project in summer 2009, a closing day was organized with research presentations in the morning and a feedback workshop in the afternoon. Unfortunately, due to a scheduling problem, only relatively few teachers and pupils attended this workshop. Nevertheless, some conclusions could be drawn and linked to comments obtained in informal discussions after the interventions.

One teacher, who expressed already after one of our interventions her astonishment how much material we have covered during a two-hour intervention, mentioned that she successfully transferred the approach to another class. From another teacher (not present) we also know that he "copied" already one unit and he invited also two other colleagues to attend an intervention done in his class. One teacher expressed her interest to have another unit for her class before she retires. Throughout the year, we got only one clearly negative feedback from a teacher (not present in the workshop). He indicated that the material presented (a hardware unit) was far too difficult for his class (lower secondary) and that it is not motivating if students get instructed in informatics without all the excellent animation tools already available. Hence, with this exception, teachers appreciated the interventions showing core concepts of informatics without being disturbed by operating a machine.

Some, but not all teachers made follow up lessons with their class. Out of them, we got written feedback from primary school children, explaining that they would like to have more *Informatik erLeben* units. One may discount this reaction, since interest might be created already by a new person appearing in class, presenting some topics which are beyond the daily routine. But referring to experience reported by Nishida et al., teaching "unplugged" concepts in classes in Japan and in Korea [11], or by Fothe reporting on role-playing for introducing algorithmic concepts [6], one might be convinced that the positive feedback is not only due to situational bias.

For us, it was important that not only the students but also the teachers reacted positively, serving as further multipliers. From more mature students, we got feedback that they particularly liked the parts with animations and the ensuing discussions. The lecturing parts, however, should not become too long. The latter observation was made by us already during the interventions. Still, some introduction is needed with any unit. Introductory parts requiring (even shallow) mathematical knowledge were not highly appreciated by the pupils. This might be due to the misconception of informatics being just use of computers. After animated search- or sorting-tasks, however, we did not sense (or hear about) problems, e.g., when computing the time it takes to execute the respective algorithm on a huge data-set.

5 Conclusion

The paper reported on a project aiming to straighten out misconceptions of pupils concerning informatics as a scientific discipline and profession and to make them curious about the challenges and merits of a technical scientific discipline. At the same time, we wanted to drag teachers away from training only computer skills instead of fundamental ideas offered by informatics.

With pupils, an intermediate assessment of the project points towards success. From an intermediate assessment, one has to conclude that success is easier reached with primary school kids than with secondary and upper secondary school students.

Concerning teachers as secondary target group, primary teachers showed the highest enthusiasm. Most of them provided wrap-up sessions after our interventions. But we have no evidence that they will adopt the approach on their own. However, due to class sequences, this option is so far not feasible. With secondary school teachers, some reported already that they tried our approach on their own with other classes. One even developed an extension to a unit.

Overall, we have to state our approach of intervening in school, getting in touch with both pupils and their teachers proved to be effective to the extent that some misconceptions were removed. For younger pupils, interest has been created, for more mature ones at least awareness has been created. Further, the approach gave us the chance to reach teachers in a way that is not feasible in formal courses of continuous education.

Acknowledgement

Funding of the project by kwf, the Carinthian Research Promotion Fund, and by the initiative *generation innovation* of bmvit, the Ministry for Traffic, Innovation and Technology, is gratefully acknowledged.

References

1. Antonitsch, P., Krainer, L., Lerchster, R., Ukowitz, M.: Kriterien der Studienwahl von Schülerinnen und Schülern unter spezieller Berücksichtiung von IT-Studiengängen an Fachhochschule und Universität; IFF-Forschungsbericht, Universität Klagenfurt, März (2007)
2. Bell, T., Witten, I.H., Fellows, M.: Computer Science Unplugged. An enrichment and extension programme for primary-aged children (5. 8. 2009),
   http://www.google.com/educators/activities/
   unpluggedTeachersDec2006.pdf

3. Bischof, E., Mittermeir, R.: Informatik erLeben: Beispiele für schülerinnen- und schüleraktivierenden Informatikunterricht, Inst. f. Informatik-Systeme, Alpen-Adria Univrsität Klagenfurt (2008)
4. Denning, P.J.: Great Principles in Computing Curricula. In: Proc. 35th SIGCSE technical symposium on Computer Science education, pp. 336–341. ACM, New York (2004)
5. ECDL Fondation, homepage, http://www.ecdl.org/publisher/index.jsp
6. Fothe, M.: Algorithmen in spielerischer Form. In: Stechert, P. (ed.) Informatische Bildung in der Wissensgesellschaft, Universitätsverlag Siegen, pp. 31–42 (2007)
7. Gallenbacher, J.: Abenteuer Informatik – IT zum anfassen von Routenplaner bis Online-Banking. Elsevier GmbH, München (2007)
8. Helmke, A., Renkl, A.: Das Münchner Aufmerksamkeitsinventar (MAI). Ein Instrument zur systematischen Verhaltensbeobachtung im Unterricht. In: Diagnostica 1992, Heft 2, pp. 130–141 (1992)
9. Kolczyk, E.: Algorithm – Fundamental Concept in Preparing Informatics Teachers. In: Mittermeir, R.T., Sysło, M.M. (eds.) ISSEP 2008. LNCS, vol. 5090, pp. 265–271. Springer, Heidelberg (2008)
10. Micheuz, P.: Zahlen, Daten und Fakten zum Informatikunterricht an den Gymnasien Österreichs. In: Körber, B. (ed.) Zukunft braucht Herkunft, Proc. INFOS 2009. GI Lecture Notes in Informatics, vol. 156, pp. 243–254. Gesellschaft für Informatik (2009)
11. Nishida, T., Idosaka, Y., Hofuku, Y., Kanemune, S., Kuno, Y.: New Methodology of Information Education with "Computer Science Unplugged". In: Mittermeir, R.T., Sysło, M.M. (eds.) ISSEP 2008. LNCS, vol. 5090, pp. 241–252. Springer, Heidelberg (2008)
12. Reiter, A.: Incorporation of Informatics in Austrian Education: The Project "Computer-Education-Society" in the School Year 1984/85. In: Mittermeir, R.T. (ed.) ISSEP 2005. LNCS, vol. 3422, pp. 4–19. Springer, Heidelberg (2005)
13. Schubert, S., Schwill, A.: Didaktik der Informatik. Spektrum Akademischer Verlag (2004)

Object-Oriented Modeling of Object-Oriented Concepts
A Case Study in Structuring an Educational Domain

Michela Pedroni and Bertrand Meyer

Chair of Software Engineering, ETH Zurich, Switzerland
{michela.pedroni,bertrand.meyer}@inf.ethz.ch

Abstract. Teaching introductory object-oriented programming presents considerable challenges. Some of these challenges are due to the intrinsic complexity of the subject matter — object-oriented concepts are tightly interrelated and appear in many combinations. The present work describes an approach to modeling educational domains and reports on the results for object-orientation. It analyzes the dependency structure of object-oriented concepts and describes the implications that the high interrelatedness of concepts has on teaching introductory programming.

1 Introduction

One of the strengths of the object-oriented mode of software development is to provide us with a set of powerful and expressive concepts, so powerful and expressive indeed that they can serve beyond their original target area — programming. These concepts, such as classes, message passing, single and multiple inheritance, were initially programming concepts; but they are in fact useful for a far more general purpose: designing systems, modeling systems, and more generally *thinking about* systems. The systems at hand are not even necessarily *software* systems: they can be human and artificial systems of many different kinds. In this work we apply the concepts to a human-centered problem: teaching. We show that it is possible and useful to take ideas originally developed for programming and apply them to the modeling of teaching and learning activities.

Partly by coincidence, the pedagogical target area — the topics for which we hope to support and improve teaching — is programming, and indeed the very form of programming whose results serve as inspiration for the teaching methods and tools: object-oriented programming. The work is then about *object-oriented techniques* for teaching *object-oriented programming*.

Teaching introductory programming is a difficult endeavor. On the side of the learner, programming is a complex activity that involves skills and mental models that many novices struggle to develop during programming courses. On the side of the instructor, teaching programming presents considerable difficulties and has been described as one of the seven grand challenges in computing education [1].

Since the mid 1990s, object-oriented programming has entered the classrooms of introductory programming courses. Many schools have since then adopted an

J. Hromkovič, R. Královič, and J. Vahrenhold (Eds.): ISSEP 2010, LNCS 5941, pp. 155–169, 2010.

"objects-first" or "objects-early" approach for their CS1 courses, and researchers as well as educators have proposed numerous tools, approaches, and strategies.

It has been asserted that for object-oriented programming "the basic concepts are tightly interrelated and cannot be easily taught and learned in isolation" [2]. This complexity is intrinsic to object-orientation and cannot be removed making it important to develop appropriate tools and processes to handle the resulting challenges.

In programming courses, it seems natural to expose students first to single programming language features (matching the first stage of Linn's "chain of cognitive accomplishments from computer programming instruction" [3]). For object-oriented programming, it is difficult to isolate single language features and to find an initial sequence of single language features (a phenomenon known as "Big Bang problem"). In addition, the tight interrelatedness of O-O concepts results in a higher number of elements to teach, since the instructor must examine not only the elementary concepts but also their possible combinations. This makes it harder to ensure that the teaching sequence meets the prerequisites at all times and that a course covers all facets of a concept.

This work describes a modeling approach and the supporting tool for modeling educational domains through their main concepts and the relations between these concepts, and its application to the educational domain of introductory programming.

2   Truc Framework

A course will never be specified as precisely and rigorously as, for example, a computer program. Still, applying modeling techniques partly imitated from software and other engineering disciplines can help meet some of the challenges of course design, in particular for object-oriented programming.

The Truc framework [4] used in this work models educational domains and identifies structural dependencies between concepts. It extends the idea of Truc (Testable, Reusable Unit of Cognition) [5] by adding two additional types of knowledge units. The final model then uses three types of knowledge units (in increasing level of granularity): *notions*, *trucs*, and *clusters*. In addition, it defines several types of relationships between the entities.

At the highest level, a *cluster* is a collection of trucs and other clusters representing a particular knowledge area. A truc belongs to exactly one cluster; the set of clusters forms a hierarchical structure in a directed acyclic graph.

At the medium level, a *truc* is "a collection of concepts, operational skills and assessment criteria" [5]. Its description follows a standardized scheme with sections on technical properties (for example, its role in the field, benefits of applying it, and a summary) and pedagogical properties (such as common confusions and sample exam questions). To help instructors check that their teaching material addresses the misconceptions of students, we have extended the original "common confusions" section [5] with *recommendations* applicable to teaching material such as slides.

The most elementary unit, *notion*, "represents a single concept or operational skill or facet of a concept" [4]. Since the key unit of granularity of the model is truc, every notion belongs to exactly one truc. A truc may have a central notion, which then bears the same name. In our example pedagogical domain, examples of notions within a "feature call" truc are: the central notion "feature call" (capturing the general idea of a method call instruction), "multi-dot feature call" (calls of the form o1.o2.o3.f), and "unqualified feature call" (method calls without an explicit target).

To capture the dependency structure of the knowledge units, the Truc framework defines two types of relations between notions. A *requires* link captures that understanding a notion requires knowing another notion. This relation is comparable to the client relationship between classes in object-oriented systems. A *refines* link expresses that a notion is a specialization of another notion; it is comparable to the inheritance mechanism in object-oriented systems. A refined notion implicitly inherits all the *requires* links from its ancestor, but may also introduce additional ones. For simplicity, the methodology prohibits *refines* links across truc boundaries.

Dependencies at the notion level contribute to dependencies at the truc level: a truc A *depends* on another truc B if any of its notions *requires* a notion of B. Since each truc contains a set of notions, the trucs and notions define a two-layered graph. The graph provides the domain model for the modeling of courses and their associated lectures as a sequence of covered notions. Figure 1 shows an extract of the **truc-notion graph**[1] for object-oriented programming. It includes the direct dependencies of truc Feature call and their notions. The textual description of an example truc is available in Appendix A.

The TrucStudio[2] [6] Pedagogical Development Environment supports the Truc approach. It automatically deduces the truc dependencies from the notion requirements; it provides a graphical representation of the domain model (such as the one produced for the truc *Feature call* shown in Figure 1) and a view of courses as diagrams. Additionally, it offers a customizable output generation mechanism to produce Web pages and ontology files, supports the analysis of transitive dependencies and cycles on notion and truc level, and reports prerequisite violation within a course.

3   Model of Object-Oriented Programming

Several articles and standards have guided the work of selecting concepts and skills that can serve as a starting point for defining the trucs of OOP. In particular, the article on "the quarks of object-oriented development" [7] identifies *inheritance, object, class, encapsulation, method, message passing, polymorphism,* and *abstraction* as "quarks". Except for *abstraction*, all of these quarks appear

---

[1] To prevent misunderstandings related to the entity type "cluster", we use the name *truc-notion graph* instead of *clustered notions graph* as found in an earlier article [4].

[2] Available at http://trucstudio.origo.ethz.ch

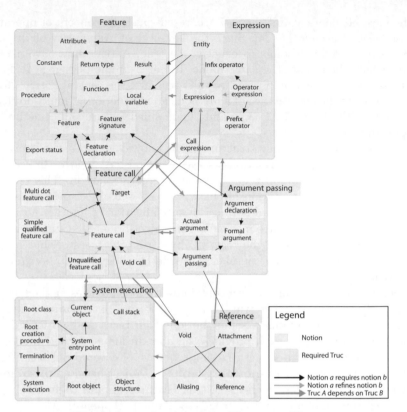

Fig. 1. Dependencies of the "Feature call" truc

as trucs (*encapsulation* under the name *information hiding* and *message passing* as *feature call*).

An experiment by Sanders et al. [8] contrasts the expert view of the quarks with the view of students who recently had studied object-oriented programming. They asked them to draw concept maps [9] that summarize their knowledge of OOP. The most commonly mentioned concepts are *class*, *method*, *instance*, *variable*, and *object*. Other commonly found concepts (implicitly or explicitly) are *data/attribute/instance variable*, *inheritance*, and *encapsulation*. The developed trucs contain all of these concepts; *instance* is integrated in the *object* truc and *data/attribute/instance variable* in the *feature* truc.

Schulte and Bennedsen [10] carried out a study in 2006 where they asked computer science teachers from high schools, colleges, and universities in various countries to rate the difficulty, relevance, and cognitive level of 28 programming topics. They refer to a set of other studies [11,12,13] that helped develop the list of topics. The topics with highest relevance are *selection and iteration*, *simple data structures*, *parameters*, *scope*, *object and class* and *syntax*. The trucs in our model cover these topics, except for *syntax*.

The ACM curricular initiative CC2001 [13] defines the body of knowledge of computer science by specifying 14 knowledge areas ranging from Discrete

Structures, Programming Fundamentals, to Social and Professional Issues. Each knowledge area contains a set of units, which hold a set of topics. It also defines six curricular models for introductory courses and proposes a syllabus and set of units for each variant. The syllabi and description of knowledge units have also guided the selection of concepts covered by trucs.

The model we have developed for object-oriented programming contains the two clusters *Object-oriented programming* and *Data structures* with 28 trucs: *Algorithm, Argument passing, Array, Assignment, Class, Conditional, Deferred class, Design by Contract, Dynamic binding, Expression, Feature, Feature call, Genericity, Hash table, Information Hiding, Inheritance, Instruction, Linked list, Loop, Multiple inheritance, Object, Object creation, Polymorphism, Primitive type, Recursion, Reference, Stack, System execution.* The trucs cover concepts ranging from imperative to object-oriented programming and simple data structures. They contain 147 notions with 196 *requires* and 39 *refines* links. These links result in 85 direct dependencies between trucs. The entire model is available as Web pages and as a TrucStudio project at http://se.ethz.ch/people/pedroni/trucs

4 Analysis of the Dependency Structure

The first part of this section analyzes the transitive dependencies and cycles as present in our developed domain model. As the domain model represents our view of object-oriented programming and is influenced by our context (in particular the programming language we use, Eiffel), we present a comparison to a model developed by another instructor using Java in the second part.

4.1 Transitive Dependencies

The analysis of the dependency structure relies on the transitive (direct and indirect) dependencies resulting from the truc-notion graph of our model for object-oriented programming. The discussion distinguishes between outgoing and incoming links. The analysis of outgoing links organizes the trucs according to the number of their dependencies (prerequisites). This gives an intuition of a truc's place in a course; trucs with many dependencies are likely to appear towards the end, while trucs with few dependencies will probably appear at the beginning. With the incoming links of trucs, the focus shifts to the number of trucs that rely on a given one. This gives an indication of a trucs' importance; if many trucs rely on it, then it is probably central to teaching programming and will reappear throughout a course. Table 1 presents the trucs grouped by their transitive outgoing dependencies: if a set of trucs shares their dependencies, then they are listed in one row.

Outgoing links. The first row of the table shows a *core group* of trucs. They constitute a minimal set of requirements for all 28 trucs appearing in the model. The core group contains nine trucs: *Argument passing, Class, Expression, Feature, Feature call, Object, Object creation, Reference,* and *System execution.* Every member of the core group depends on itself and on all other members. This

Table 1. Overview of transitive truc dependencies

Truc \ Prerequisite	Core group	Algorithm	Array	Assignment	Conditional	Deferred class	DbC	Dyn. binding	Genericity	Hash table	Inf. hiding	Inheritance	Instruction	Linked list	Loop	Multiple inh.	Polymorph.	Prim. type	Recursion	Stack	# Outgoing links
Core group: Argument passing, Class, Expression, Feature, Feature call, Object, Object creation, Reference, System execution / Assignment, Inheritance, Primitive type	x																				9
Deferred class, Genericity, Multiple inh.	x											x									10
Conditional, Instruction, Loop / Algorithm	x			x	x								x		x			x			14
Design by Contract	x											x						x			11
Polymorphism	x			x					x			x									12
Dynamic binding	x			x					x			x					x				13
Information hiding	x			x				x	x			x					x				14
Array, Linked list	x	x		x	x				x			x	x		x			x			17
Hash table	x	x	x	x	x				x			x	x		x			x			18
Stack	x	x	x	x	x				x			x	x	x	x			x			19
Recursion	x	x	x	x	x				x			x	x	x	x			x		x	20
# Incoming links	28	5	3	12	9	0	0	1	8	0	0	12	9	2	9	0	2	10	0	1	

is an indication for cycles in the domain model (see 4.2). The trucs *Assignment*, *Inheritance* and *Primitive type* share the dependencies of the core group. They are not part of the core group, because they are not all mutually dependent.

The second set of trucs with cyclic dependencies consists of *Conditional*, *Loop*, and *Instruction*. They are recursively dependent on each other. Additionally to the core trucs, they depend on *Assignment* and *Primitive type*. *Algorithm* has the same dependencies, but does not recursively depend on itself. This group mostly contains trucs associated to imperative programming.

All remaining trucs require *Inheritance* besides the nine core trucs. This is the only supplemental requirement for *Deferred class*, *Genericity*, and *Multiple inheritance*. *Design by Contract* additionally relies on *Primitive type*, while *Polymorphism* requires *Assignment* and *Genericity* in addition to *Inheritance* and the core trucs. *Dynamic binding* depends on *Polymorphism* and thus includes all its dependencies; similarly, *Information hiding* relies on *Dynamic binding* and shares all its requirements. This group mostly contains advanced object-oriented concepts related to inheritance.

The trucs representing knowledge about data structures combine the dependencies of the imperative programming group with some of the object-oriented group. *Linked list* and *Array*, for example, depend on *Algorithm, Conditional, Instruction, Loop,* and *Primitive type*, as well as on *Inheritance* and *Genericity. Hash table* additionally requires *Array; Stack* requires both *Array* and *Linked list*; and *Recursion* depends on *Stack*.

Incoming links. The nine core trucs are a prerequisite for all the trucs of the domain model. This makes them fundamental for teaching object-oriented programming. The second group containing *Assignment, Inheritance* and *Primitive type* are required by 12 respectively 10 other trucs (almost half of all trucs) and *Genericity* is a requirement for eight trucs. The second cyclic group containing *Conditional, Loop, Instruction,* and *Algorithm* provides a basis for nine respectively five trucs. *Polymorphism, Dynamic binding, Array, Linked list,* and *Stack* are prerequisite to one to three trucs. The trucs *Deferred class, Multiple inheritance, Design by Contract, Information hiding, Hash table* and *Recursion* do not appear as a requirement for any truc in the model.

Transitive dependencies of notions. The transitive dependencies between notions exhibit characteristics similar to those of the trucs. Ten notions, out of 147, form a core group such that all notions in the model transitively require them. The core group contains the notions *Argument declaration, Class, Feature, Feature declaration, Feature signature, Formal argument, Generating class, Instance, Object,* and *Type*. Additionally, over half of all notions in the model transitively require the notion *Expression*. 55 notions are not needed by any other notions.

4.2 Cycles

On the notion level, the domain model exhibits five circular dependencies, of which three involve the truc *Argument passing*. One of these cycles is a mutual dependency between the notions *Argument declaration* and *Feature signature*; another cycle consists of the notions *Argument passing, Actual argument,* and *Feature call*; and the third cycle contains the notions *Formal argument, Type, Class, Feature, Feature declaration, Feature signature,* and *Argument declaration*. The fourth cycle on the notion level involves the trucs *Class* and *Object* and illustrates their close interrelatedness via a path through the notions *Class, Object, Instance,* and *Generating class,* back to notion *Class*. The fifth cycle shows the close connection between *Function* and *Result*.

As indicated in 4.1, two groups of trucs contain cycles in their lists of dependencies. Figure 2(a) shows an extract of the graph with the direct dependencies of the core trucs *Argument passing, Feature call, Feature, Class, Expression, Object creation, Object, Reference,* and *System execution*. This subgraph exhibits high interrelatedness between its concepts; in particular, there are multiple pairs of mutual dependencies (such as between *Class* and *Object*, and *Feature call* and *Expression*). Figure 2(b) shows the second group of trucs with mutual dependencies that connect *Instruction* to, separately, *Conditional* and *Loop*.

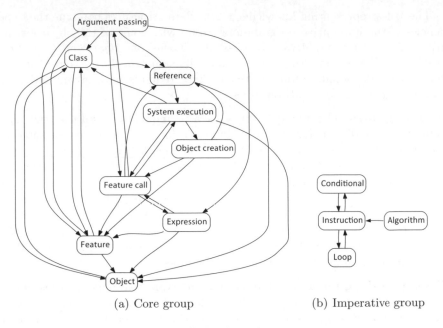

(a) Core group   (b) Imperative group

Fig. 2. Cyclic dependencies

4.3   Comparison with Another Model

Our use of Eiffel to teach programming has some bearing on the model of object-oriented programming. The choice of trucs and notions, the relationships between notions, and the descriptions of the trucs reflect this particular choice. The model may not, as a result, reflect a generally accepted image of object-orientation and it is not the only form of object-oriented programming.

To find out which properties, in particular cyclic and transitive dependencies, might be artifacts of our course's choices, we asked another instructor teaching introductory object-oriented programming with Java to model parts of his teaching. His domain model includes the entities required to represent the first three lectures of his introductory Java course. It contains a cluster *Programming* with the three trucs *Programming language*, *Memory management*, and *Program* and a second cluster *Object-orientation in Java* with the 12 trucs *Data type*, *Object*, *Method*, *Variable*, *Polymorphism*, *Compilation unit*, *Instruction*, *Expression*, *Access modifier*, *Conditional*, *Loop*, and *Identifier*. The trucs contain 67 notions with 19 *refines* links and 33 *requires* links. Appendix B shows the model as a truc-notion graph. The notion dependencies are incomplete.

A comparison of this model to ours shows that they cover similar notions, but their distribution amongst trucs varies. For example, the Java truc *Method* combines sets of notions from our trucs *Feature* and *Argument passing* with single notions of *System execution*. A similar pattern is visible for other trucs, such as the Java truc *Object*, which subsumes our truc *Object* and includes single notions found in our trucs *Feature call*, *Reference*, and *Object creation*. Our model has no truc conforming to the Java truc *Access modifier*, but its notions are integrated in truc *Feature*. Additionally, certain trucs have different names although covering similar notions. For example, the Java truc *Polymorphism* conforms to our truc *Inheritance* and *Compilation unit* conforms to *Class*.

Analysis of the transitive dependencies of the Java model results in a core group containing the trucs *Data type*, *Method*, *Object*, and *Variable*. These trucs are suppliers for about half of the trucs in the model. In particular, they are a prerequisite to themselves and to the trucs *Instruction*, *Polymorphism*, *Memory management*, and *Access modifier*.

There are no cycles on the notion level in this model, but four cycles exist at the truc level: *Method* and *Data type* as well as *Object* and *Variable* are mutually dependent; additionally, there is one cycle containing the trucs *Method*, *Variable*, and *Object*, and one containing *Method*, *Data type*, and *Object*.

A few differences exist between the two models. In the Java model, truc *Compilation unit* containing the *Class* notion is not part of the cyclic group, while in our model *Class* is part of the core group. On the other hand, the truc *Data type* conforming to our truc *Primitive type* is part of the core group. This is due to the notion *Command line argument* in the *Method* truc requesting the notion *Array* of truc *Data type*.

Another difference is that most of the other parts of the model are unconnected. This is probably due to the incomplete nature of the model.

The Java model exhibits similar characteristics with respect to transitive dependencies and cycles as ours. The most striking similarity is the existence of a group of core trucs that are mutually dependent and that are fundamental for a large portion of the remaining trucs. This suggests that our model has a broader reach than just our course.

5   Implications for Teaching

Instructors face many challenges when designing courses or textbooks. Besides pedagogical finesse for presenting material adapted to students' skills and interests, they must demonstrate the ability to structure the material in a sound sequence. This task is particularly difficult if the domain of teaching is object-oriented programming, due to the high interrelatedness of its concepts [14,15], what Caspersen calls "one of the most challenging inherent complexities of object-orientation" [2, p. 78]. It complicates finding a starting point where no prerequisites are necessary, and raises the challenges of how to cover the entire subject area and how to order the concepts without prerequisite violations.

Analysis of the truc and notion graphs in this article confirms Caspersen's observation, but narrows it down to a core group of nine closely interrelated concepts. This group contains the trucs *Argument passing*, *Class*, *Expression*, *Feature*, *Feature call*, *Object*, *Object creation*, *Reference*, and *System execution*. Their transitive and recursive dependencies show that they belong in the initial phase of an objects-first course.

On the notion level, the ten notions *Argument declaration*, *Class*, *Feature*, *Feature declaration*, *Feature signature*, *Formal argument*, *Generating class*, *Instance*,

Object, and *Type* have similar properties with respect to their dependencies as the nine core trucs. With the exception of *Argument passing* and *Formal argument*, which can be omitted if only features without arguments are covered, it seems necessary to introduce them together.

The circular dependencies in the truc and notion model indicate that teaching object-oriented programming requires a spiral model; "A curriculum as it develops should revisit these basic ideas repeatedly, building upon them until the student has grasped the full formal apparatus that goes with them" [16]. The detected cycles also confirm the existence of the "Big Bang problem" [2].

The Inverted Curriculum [17,18] approaches the Big Bang problem by using a large body of supporting software that, through information hiding and inheritance, allows the initial examples and exercises used in class to rely on advanced mechanisms without introducing them formally yet. For example, it first introduces feature calls on predefined objects inherited from an ancestor class. The difficulty of the Inverted Curriculum approach is that the preparation of the software framework and all examples and exercises needs to happen before the students receive the software. This produces the need for more planning and may lead to a restricted set of possible exercises.

Another approach to handling the Big Bang problem is to use an "example-driven" approach, where the "progression in a course is defined by increasing complexity of class models rather than being dictated by a bottom-up ordering of language constructs" [19]. For introducing association, for example, this approach first uses recursive 0..1 association (a `PERSON` class having an attribute `married_to`), then it covers 0..* associations (extending `PERSON` with attribute `friends`), and finally it shows associations between different classes. The language constructs and concepts required for understanding the examples are introduced when they are needed (such as collections for recursive 0..* associations).

The concept interrelatedness also makes it difficult to ensure that an existing course is compatible with the prerequisites and covers all the concepts. We have modeled our Introduction to Programming course using TrucStudio and detected one critical prerequisite violation (the *Result* entity is used before introducing *Function*) and five notions missing in the course slides (*Multi-dot feature call, Manifest constant, Constant, Precursor*, and *Polymorphic creation*).

The complexity of O-O concepts also leads to misconceptions in students' understanding when they first learn about them. The created trucs and their common confusions sections help check whether the teaching material addresses these misconceptions. In a first analysis of our teaching material, we have found that it addresses only a small part of the misconceptions. The conjecture is that this phenomenon is not specific to our material, but more general: although a large body of studies on novices' misconceptions is available, it rarely influences actual teaching.

6   Conclusions

This article has presented a modeling approach for educational domains and reported on its application to object-oriented programming. The approach uses knowledge units at three levels: clusters (describing knowledge areas), trucs (describing skills and concepts following from a central idea), and notions (describing single facets of a concept). It also includes links between the entities to capture the dependency structure of a domain.

The resulting domain model for object-oriented programming consists of 28 trucs and 147 notions. The analysis of the dependency structure confirms that the basic object-oriented concepts are tightly interrelated. It identifies a core group of trucs containing *Argument passing, Class, Expression, Feature, Feature call, Object, Object creation, Reference,* and *System execution.* The core trucs are mutually dependent; all other trucs in the model rely on them. The core group of trucs indicates that the associated concepts are mostly responsible for the Big Bang problem — the problem of finding a proper order of introduction for the basic concepts of object-oriented programming.

The developed domain model is influenced by our context (in particular by the programming language we use, Eiffel). To ensure that the findings from our domain model apply to a more general context, we have analyzed a second model of object-oriented programming developed by an instructor using Java for his introductory programming course. In spite of differences in the trucs and in links between notions, his model also contains a mutually dependent core group, on which approximately half of all trucs rely. This indicates that the model for Eiffel is similar to the one for Java and that the results are likely to apply to yet other settings. In the future, we would like to develop a more complete model for Java (possibly based on the Eiffel trucs) and to investigate whether mechanisms from mathematics and computing can help teach tightly coupled notions (as proposed by a reviewer of this article).

Acknowledgements. We thank D. Herding from RWTH Aachen for testing TrucStudio and providing a partial model of object-oriented programming in Java. We are grateful to the anonymous referees for many useful comments.

References

1. McGettrick, A., Boyle, R., Ibbett, R., Lloyd, J., Lovegrove, G., Mander, K.: Grand Challenges in Computing: Education A Summary. The Computer Journal 48(1), 42–48 (2005)
2. Bennedsen, J., Caspersen, M.E., Kölling, M.: Reflections on the Teaching of Programming. Springer, Heidelberg (2008)
3. Linn, M.C., Dalbey, J.: Cognitive consequences of programming instruction. Studying the Novice Programmer, pp. 57–81. Lawrence Erlbaum Associates, Mahwah (1989)
4. Pedroni, M., Oriol, M., Meyer, B.: A framework for describing and comparing courses and curricula. SIGCSE Bull. 39(3), 131–135 (2007)
5. Meyer, B.: Testable, reusable units of cognition. IEEE Computer 39(4), 20–24 (2006)

6. Pedroni, M., Oriol, M., Meyer, B., Albonico, E., Angerer, L.: Course management with TrucStudio. In: ITiCSE 2008: Proceedings of the 13th annual conference on Innovation and technology in computer science education, pp. 260–264. ACM, New York (2008)
7. Armstrong, D.J.: The quarks of object-oriented development. Commun. ACM 49(2), 123–128 (2006)
8. Sanders, K., Boustedt, J., Eckerdal, A., McCartney, R., Moström, J.E., Thomas, L., Zander, C.: Student understanding of object-oriented programming as expressed in concept maps. SIGCSE Bull. 40(1), 332–336 (2008)
9. Novak, J.D., Cañas, A.J.: The theory underlying concept maps and how to construct them. Technical report, IHMC CmapTools, Florida Institute for Human and Machine Cognition (January 2006)
10. Schulte, C., Bennedsen, J.: What do teachers teach in introductory programming? In: ICER 2006: Proceedings of the second international workshop on Computing education research, pp. 17–28. ACM, New York (2006)
11. Milne, I., Rowe, G.: Difficulties in learning and teaching programming - views of students and tutors. Education and Information Technologies 7(1), 55–66 (2002)
12. Dale, N.: Content and emphasis in CS1. SIGCSE Bull. 37(4), 69–73 (2005)
13. The Joint Task Force on Computing Curricula: Computing Curricula 2001 (final report). Technical report, ACM and IEEE (December 2001), http://www.acm.org/sigcse/cc2001
14. Gries, D.: A principled approach to teaching OO first. SIGCSE Bull. 40(1), 31–35 (2008)
15. Shultz, G.: Using a restricted subset of Java in the first part of CS1. J. Comput. Small Coll. 23(1), 212–218 (2007)
16. Schwill, A.: Fundamental ideas in computer science. Bulletin European Association for Theoretical Computer Science 53, 274–295 (1994)
17. Meyer, B.: The outside-in method of teaching introductory programming. In: Broy, M., Zamulin, A.V. (eds.) PSI 2003. LNCS, vol. 2890, pp. 66–78. Springer, Heidelberg (2004)
18. Pedroni, M., Meyer, B.: The inverted curriculum in practice. In: SIGCSE 2006: Proceedings of the 37th SIGCSE technical symposium on Computer science education, pp. 481–485. ACM Press, New York (2006)
19. Bennedsen, J., Caspersen, M.: Model-Driven Programming. In: Reections on the Teaching of Programming, pp. 116–129. Springer, Heidelberg (2008)

Appendix A – An Example Truc: Feature Call

Alternative names	Method invocation, Message passing
Dependencies	Feature, Object, Argument
Notions	Feature call, Multi dot feature call, Simple qualified feature call, Target, Unqualified feature call, Void call

Summary. Feature call is the mechanism of applying a feature to a target object [Meyer, 2009]. The target may be explicitly defined through an expression, which at run time will be attached to a certain object. If no explicit target is given, the target is the Current object. Feature calls may contain arguments.

Role. Feature call is the "basic mechanism of object-oriented computation". In an object-oriented software system, all computation is achieved by calling features on objects and no software element will ever be executed except as part of a feature call [Meyer, 1997].

Applicability. Need to modify an object or access object data or state.

Benefits.

- Fundamental to create running programs.
- Favors reuse of code by outsourcing a set of instructions into a feature and replacing them by a single feature call.

Pitfalls. Using non-pure queries (queries that change the state of an object) in feature calls may produce side effects. Side effects may lead to mysterious failures that are difficult to locate and fix.

Common confusions.

- **Static-text execution.** Many novices have an incorrect model of control flow for feature calls. It is difficult for them to understand that a feature call results in the suspension of the calling feature (caller), a transfer of the execution control and sometimes data to a *new and unique specimen* of the called feature (callee), then a transfer of control and data back to the caller after the callee has finished. This knowledge is especially important when recursion is introduced. [George, 2000]
- **Availability of features.** "Calling a non-existent feature" seems a recurring mistake [Ng Cheong Vee et al., 2006]. Writing feature calls demands thorough knowledge of what features are available for an entity. This requires looking up the type of the entity and the sufficiently exported features.
- **Wrong target.** Various errors in connection with feature calls originate from specifying wrong targets. For example, students use an unqualified feature call where a qualified feature call is necessary, or they explicitly specify the target to be Current for features that are only available for unqualified feature calls. [Ng Cheong Vee et al., 2006]
- **Object state.** Novices have difficulties in understanding that feature call instructions modify object state. [Ragonis and Ben-Ari, 2005]

- **Query as command.** It seems difficult for novice programmers to distinguish between queries and commands for feature call instructions. In certain programming languages it is possible to use a query (with a resulting value/object) as a statement, but the result is then lost. This is a common mistake. [Hristova et al., 2003]

☐ Examples show control flow of feature calls.
☐ Examples include lookup of available features.
☐ Examples include feature calls on composite targets.
☐ Examples show when qualified or unqualified calls are appropriate.
☐ Examples include feature calls that modify object state.
☐ Examples show invalid (or unwanted) use of expression as instruction.

Sample questions. Consider a class *WORD* that has an attribute *word*: *STRING*, a procedure *set_word(s*: *STRING)*, and a procedure *print* that displays the word on a console. It also has a function *substring(i*, *j*: *INTEGER)*: *WORD*, which returns a new *WORD* object containing the part of the original word defined through the indices *i* and *j*. Given is an entity *w*: *WORD*. Write an instruction that sets the *word* entity of *w* to "summertime". Then write an instruction that uses the *substring* function to extract "time" from *w*.

Feature Call Bibliography

[George, 2000] George, C.E.: Erosi – visualising recursion and discovering new errors. SIGCSE Bull. 32(1), 305–309 (2000)

[Hristova et al., 2003] Hristova, M., Misra, A., Rutter, M., Mercuri, R.: Identifying and correcting Java programming errors for introductory computer science students. SIGCSE Bull. 35(1), 153–156 (2003)

[Meyer, 1997] Meyer, B.: Object-Oriented Software Construction, 2nd edn. Prentice-Hall, Englewood Cliffs (1997)

[Meyer, 2009] Meyer, B.: Touch of class: Learning to program well with objects and contracts. Springer, Heidelberg (2009)

[Ng Cheong Vee et al., 2006] Ng Cheong Vee, M.-H., Meyer, B., Mannock, K.L.: Empirical study of novice errors and error paths in object-oriented programming. In: 7th Annual HEA-ICS conference, Dublin, Ireland (2006)

[Ragonis and Ben-Ari, 2005] Ragonis, N., Ben-Ari, M.: On understanding the statics and dynamics of object-oriented programs. SIGCSE Bull. 37(1), 226–230 (2005)

Appendix B – Java Model of Object-Oriented Programming

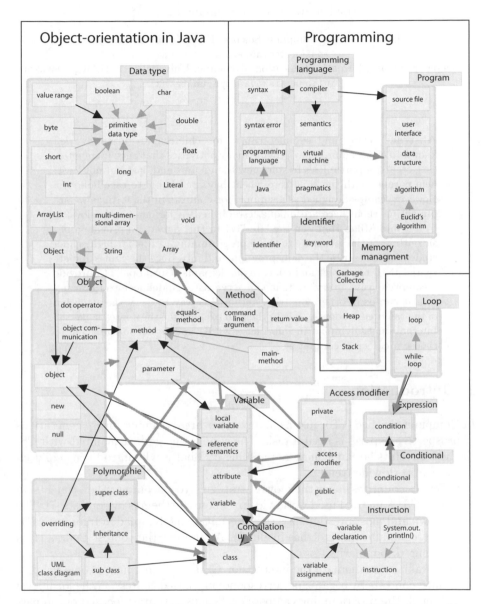

Fig. 3. Truc-notion graph of Introduction to Programming at RWTH Aachen (done by D. Herding, German terms translated into English)

Programming Camps: Letting Children Discover the Computer Science⋆

Monika Steinová[1] and Juliana Šišková[2]

[1] Department of Computer Science, ETH Zurich, Switzerland
monika.steinova@inf.ethz.ch
[2] Department of Informatics Education, Comenius University, Bratislava, Slovakia
juliana.siskova@fmph.uniba.sk

Abstract. Daily and summer camps for children are certainly well known to every adult. They provide many activities that encourage interests, create opportunities to start new friendships and, last but not least, help parents with the child care when they are unable to take holidays. However, camps designed as a support for contests in the form as we are presenting them here are unique, established in former Czechoslovakia by Vít and Milan Hejný in the late 70s. In this paper, we provide a basic outline of the structure of these programming camps for high school students, we describe the daily program, and introduce the games and activities used to attract and educate participants. Our hope is to provide a comprehensive overview that can serve as a guideline for organizing similar camps in other countries, and in this way to draw the attention of the talented students, to help them to prepare for the contests like the International Olympiad in Informatics (IOI) and to encourage them to choose a career in computing.

1 Introduction

In Computer Science (CS), just as in any other area, training future generations of prospective experts is a vital task. The sooner one gets acquainted with CS, the better results he can achieve in pursuing it. That is why introducing children to CS is a worldwide trend.

Different countries have their own strategies to encourage the students in exploring CS. A method used worldwide to encourage them is by contests [7]. The best known and most prestigious is the International Olympiad in Informatics [1] (and the regional contests it has inspired). At present the contest Bebras [4] is also gaining in popularity.

In Slovakia, we stimulate the interest for CS and train young people for programming contests by organizing correspondence based contest on the local level and through the programming camps. For various activities focused on preparing the students in Slovakia, see Forišek, Winczer [5]. In this paper, we present our experience in preparing the camps, not only by discussing the structure and

---

⋆ This work was partially supported by FILEP grant 351 of ETH Zurich.

J. Hromkovič, R. Královič, and J. Vahrenhold (Eds.): ISSEP 2010, LNCS 5941, pp. 170–181, 2010.

activities of the camp, but also the reasons why and the ways how it helps to teach children about the CS, develop their social skills and hunger for knowledge.

In Section 2, we outline the basic information about our scientific competition camp; in Section 3, we present the schedule of the camp. Section 4 and Section 5 are devoted to the camp's educational and social activities, respectively. Some statistics of popularity of the camp games and activities with participants are presented in Section 6. The paper is concluded in Section 7.

2   Overview

Scientific contest camps for talented youth have a long tradition and respected standing in Slovakia. The most elementary camps are oriented towards primary school students from the age of 12 years and are focused mainly on mathematics. Their objective is to popularize the field and to provide impulses beyond the scope of the school curricula.

The Correspondence Seminar in Programming (KSP) is a correspondence contest concerned with algorithmic programming with more than 25 years of tradition, in which secondary school students regularly receive letters with contest tasks. The target group of the contest are young people up to 19 years, i.e. the age when young people enter universities in Slovakia. It covers mainly the field of algorithm design through problem solving. To solve the contest problems, the participants must show that they are capable of implementation, clear explanation of the underlying ideas and reasoning about the complexity of the solution. The organizers of KSP are usually former participants that became students of Computer Science at Comenius University in Bratislava and they continue to run KSP on the voluntary basis. This is the case of both authors of this article as well.

The competition consists of four correspondence series per year with 10 tasks each. Participants send the written solutions that are returned back corrected together with sample solutions and the ranking list. The contest is handled by the postal service (recently, emails became acceptable, too). Therefore, the camps offer the only opportunity for the participants as well as the organizers to get to know each other personally.

The reward for the best 32 participants of KSP is the 6-day long programming camp organized twice per year (after a period of two correspondence series) by 10 organizers from KSP. The age of the participants varies and is usually between 15 and 19 years. A significant number of participants comes from schools without additional lectures in programming and/or mathematics. Female students are encouraged to participate in KSP. The best 4 female participants are invited to the camp, regardless of their overall ranking.

Camp location. The camp is located in a camp resort with catering in a remote area, usually without internet access. The location is selected with a purpose of relieving the participants from the influence of modern technologies and letting them instead spend their time socializing. It also allows the organizers to prepare activities which the participants can hardly experience elsewhere.

The camp team competition. At the beginning of the camp, participants are divided into 4 teams that are then competing with each other during the entire camp. The competition is more or less symbolical and the final award is only a cake for each team. The only individual competition is a series of mental exercise (see Section 4.5 for details) which is evaluated separately at the end of the camp.

The underlying benefit of the team structure is that the team members are learning from each other while answering to the impulses in the games and activities. This corresponds with Vygotsky's principles of how children learn being tutored by older children [8]. Moreover, the team structure allows the introverts to feel safe in the team whereas more extrovert people gain the space for self-realization.

Storyline. The uniqueness of each camp is achieved by creating its storyline with the usual themes include fantasy (wizards, spells), science fiction (space-ships and aliens), history (knights, prehistorical), literary tales (Harry Potter, Musketeers) or modern adventure (journalists, spies, detectives). During the camp week, the storyline gradually uncovers through a number of theme games, activities and by other means such as regular role-playing within the scope of the story, daily "printed" camp newspapers, appearance of the messages on the camp pin-board or appearance of special items in the camp area. The storyline develops until the last day, when, in a concluding main game, a final common objective has to be achieved (save the Earth/princess, return from the deserted island, etc.). The storyline gives a very special dint to every camp and makes the participants return home with positive memories.

3   The Weekly Schedule of the Camp

The overview of the weekly schedule of the camp is shown in Figure 1. For detailed description of particular activities see Sections 4 and 5.

The camp start-up games are usually placed in the schedule around the time of the first dinner. During the first part of these games, participants and organizers are introduced and a game is played to help them remember the names of each other. The second part utilizes the distribution into the teams. The teams need to be well-balanced with respect to their physical prowess and CS skills of their members.

The "Grand Prix", i.e. the main concluding game, is a climactic point of the camp storyline. Usually, it is divided into several parts that resemble the games from the social program section (see Section 5 for details).

The overall evaluation of the team competition, series of mental exercise and also the awarding of the best KSP-solvers takes place in a final closing ceremony.

The schedule of the camp might be re-arranged when a special game or activity (e.g., a whole-day hiking trip) from Section 5 is planned. Possible sports games between lectures, the period of relaxation after the lunch and the short breaks between the timeslots in the schedule are not depicted in the Figure 1 for clarity of presentation and are variable in accordance with the needs of the participants.

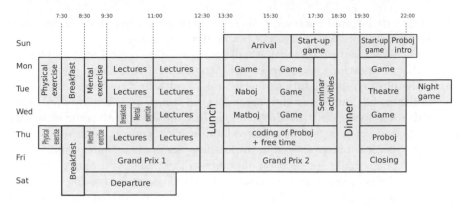

Fig. 1. The week overview of a camp schedule

4   Educational Program

In this chapter, we provide an overview of the educational activities of the camp. Besides lectures and seminars, as described in Section 4.1, these activities consist of various contests which are described in more detail in Sections 4.2 to 4.5.

4.1   Lectures and Seminar Activities

Lectures and seminar activities are the essential part of CS-teaching in the camp. They are arranged by the organizers of the camp who are all university students of CS. The fact that the lecturer was once a participant himself gives him/her a better perspective for detection and solving of the problems connected with the comprehension of the topic by the participants. The lecturing competence of the less-experienced lecturers is mitigated by careful selection of suitable topics and by the supervision of the older students from the KSP organization board. No two camps are organized by the same set of people which brings uniqueness and diversity into the content as well as into the overall realization.

The length of a lecture ranges from 45 minutes up to one and a half hour. There are usually two parallel lectures – one intended for the beginners and one for the advanced participants – and they usually take place in the morning. The content of the lecture is prepared so that the people at different levels of knowledge are able to follow it. The lecturer is ready to provide extra information if the lecture is progressing better than expected. The topics of the lectures vary from CS on the theoretical level to the presentation of the newest trends, multimedia and technologies.

In contrast to the lectures, the seminar activity is shorter, working with a small number of people only. Out of five (or more) parallel seminars at various levels of difficulty, every participant selects and attends one seminar that is a series of 3 or 4 one-hour meetings per camp. Since the number of participants of a seminar is usually limited to eight people, the presentation of the topic is very individual and allows for the use of different forms of presentation methods including short internal contests, games, etc. The topics discussed during these sessions are more oriented towards CS and algorithm design.

Some beginner lecture topics that will prove to be successful are: *simple complexity* (big O-notation), *data structures* (array, linked list, stack, queue), *basic graphs* (definition, representation, depth- and breadth-first search), *sorting* (sorting algorithms like quick sort and merge sort), *dynamic programming* (introduction with some examples), *greedy technique* (introduction with some examples) and *divide and conquer technique* (introduction with some examples).

Other possible topics discussed in the lectures/seminars include *computational geometry* (convex hulls, localization of points in the plane, sweeping technique), *complexity theory* (amortized complexity, Master theorem for solving recurrence equations), *string algorithms* (Knuth-Morris-Pratt and pattern matching, suffix trees), *cryptology* (protocols, digital signatures, time stamps, RSA), *coding theory and compression* (Huffman, Fano, Shannon codes, compression in bzip2), *formal languages* (finite automata, parsing, nondeterminism, Turing machines, NP-completeness), *graphs* (coloring, Euler tour, spanning trees, bridges and articulations, shortest paths, topological sort), *data structures* (heap, binary search trees, interval trees, trie, hashing), *mathematics* (Euclid's algorithm, game theory), *popular software* (TeX, Linux, various programming languages, viruses), *computer architecture* (file systems, operating systems, cache), *ACM training* and many more.

4.2 Proboj – The Programming Competition

Proboj[1] is a team programming contest spanning almost the whole period of the camp. It is based on a simple multiplayer computer game that several organizers develop for the camp. During the camp, each team develops a player program. In the end, the final versions of these programs are launched simultaneously, competing against each other. The organizers prepare the Proboj's environment, i.e. the interface via which players obtain information about the state of the game, whereas teams have to develop a kind of artificial intelligence that is going to play on their behalf. Proboj is graphically very attractive and the interface used to control a player is simple – usually a single procedure is executed as a move of the player. During the week, participants often have a textual version of the game only, but then the tournament is presented with all the glamor – intro with titles of producers, supporting music and then several games in nicely rendered environment.

Since there are many different ways in which members of the team can contribute, the contest is very appealing to the participants. Usually, several people analyze strategies and possible reactions to the opponent moves. Then, a few selected members of the team transform the ideas into the code. Here, the participants can experience teamwork on a project that includes many different aspects of development.

The game rules are introduced to the participants on the day of their arrival allowing them to discuss Proboj during their free time. The afternoon of the day

---

[1] Abbreviation for "the programming fight" in Slovak.

before the last day is devoted to the development of the player programs and then, in the evening, the tournament of the final versions of players is staged.

In the previous years, several different types of Proboj were used. The most remarkable type of the Proboj programs is the turn-based strategy, where, given a map, the player has to collect objects by employing units within the game. Other types include card games (Hearts, Poker), board games (Clue, Risk, Stratego) and real-time games (car racing, tennis, ice-hockey).

4.3   Naboj – The Mathematical Team Contest

The problems are usually derived from various areas of mathematics and CS. This type of contest stimulates competitiveness and allows the participants to match their skills, that is why it is very popular. It is not physically demanding and therefore suitable as the afternoon program.

The competition usually lasts one hour, and a further half of an hour is reserved for the discussion of the solutions. Out of about 30 tasks of increasing difficulty the team always has an access to 5 tasks at one time. Each time the team solves a problem, it gets a new one. Usually the rules of the game allow the team to drop a task and get another instead. The ratio between the mathematical and the CS tasks is approximately 2:1.

Example 1. Some problems used in Nabojs include:

1. Find the smallest positive number x, such that x modulo 47 = 5 and x modulo 42 = 14.
2. You are given a program with task description. Replace one character to make the program run correctly.
3. Prove: $\forall n \in \mathbb{N}: \frac{1}{n+1} + \frac{1}{n+2} + \cdots + \frac{1}{2n} > \frac{13}{24}$
4. Decide on the validity of the following statements:
   (a) The complement of a disconnected graph is a connected graph.
   (b) The complement of a connected graph is a disconnected graph.
5. Prove or disprove: A graph is connected if and only if its edge-graph is connected.

Solving as many tasks as possible does not have to be the main goal of Naboj. The competition is often covered by a guise of a game where solving a task is reflected into one or several moves in a strategic game, revealing nodes while searching a graph, gaining clues to a Zebra puzzle[2] [3] or a Smullyan-like logic problem [6]. The covering game has to be well-balanced so that the more tasks a team solves the more points it gains. However, it should be designed in such a way that devising a good strategy can affect the overall results slightly.

---

[2] The Zebra puzzle consists of N people, several different types of sets of N items and a number of statements about the relations of these people and these items. The task is to assign one item of each type to every person so that the statements given hold and so that none of two people have the same item of one type.

The following competencies are developed during Naboj.

Brain-tingling. Mathematical and CS problems introduce the participants to the enjoyment of learning and discovering own potential and the thrill of finding the answer.

Learning by observation. Participants have different skill levels in solving the problems and different interests in topics. Therefore, the team cooperation causes participants to observe the ways of solving the tasks as done by other team members. The discussion and the sample solutions provided after the contest are another example of learning by observation.

Task management. Given 5 tasks at one moment, the team has to manage its tasks and distribute them among its eight members according to their abilities.

Skill estimation. Due to the number of tasks and the covering game, most of the tasks are being solved by one member of the team only. Therefore, the right estimation of skills is important and overestimating is expensive – it costs time and also points.

Development of strategies. Besides the need of a strategy within the covering game, the solving of the tasks opens the room for tactics. Participants are aware that the difficulty of the tasks present is increasing and that once a task is given up it cannot be solved later.

4.4   Matboj – The Game Tournament

Matboj[3] is a tournament in a given combinatorial game (e.g., variation of the NIM) or board game. The game used for the tournament is usually a two-player game, preferably with complete information; it should take only a few minutes to play, and the optimal strategy needs to be complex enough so that the players cannot detect it easily.

The tournament proceeds as follows: After the explanation of the rules of a particular game, each team has up to 30 minutes to devise a strategy and play several sample games. Then the four teams play the all-play-all tournament, i.e. the members of the two opposing teams form pairs that are playing two rounds of the game – each player starting once. The winning player receives 2 points for his/her team, in the case of a draw, both players receive 1 point. After the tournament, all points gained by a team are summed-up and the final ranking of the teams is determined.

4.5   Mental Exercise

The program of the standard camp day begins with a session of mental exercise after the breakfast. This is a regular activity where participants warm-up their brains before the lectures by solving simple algorithmic tasks individually, using pen and paper only. The series of mental exercise is the only individual competition of the camp. The best solvers of the whole week are awarded at the end of the camp.

---

[3] Abbreviation for "math fight" in Slovak.

Each session of the mental exercise consists of two algorithmic tasks. The tasks are a lighter version of the KSP tasks. They are simplified to the utmost level and given without a story just as pure formulations of problems. The participants are asked to solve both tasks in 30 minutes. Their solution should contain their core idea, pseudocode and discussion of the complexity. Mental exercise may contain tasks where knowledge gained in the beginner lectures is applied.

Example 2. Some tasks used in the mental exercise.

- Given a two-dimensional array of 0s and 1s, write a program that finds the largest square containing only 1s.
- Given two sorted arrays of different integers, write a program that computes the median[4] of the elements of both arrays.
- Given an array of integers, find the longest subsequence with the largest arithmetic average.
- Write a program that calculates the smallest number of trails necessary to draw a given graph. Each edge of the graph must belong to exactly one of the trails. (A trail is a sequence of distinct edges with each connected to the preceding one.)

5   Social Activities

Since the stay in the camp is a reward for the participants of KSP, the camp program cannot consist solely of teaching and learning about new topics. Instead of making the camp schedule school-like, the educational program is interlaid with several other activities, the most significant representatives of which are mentioned below. Except for the academy, the hiking trip and the fighting games, all other games are carried out on the team basis.

The games and activities are intended to be an enjoyable part of the camp for all the participants – by simple being a game and also by the addition of side games that can affect the course of the large game. These side small games are diverse so that each member of the team can select the activity that suits him/her best. While playing these games, participants are confronted with the competencies pointed out in the text.

Academy. Academy is an activity requiring more time, consisting of several series of lectures that imitate the university environment. The organizers of the camp prepare several mock-lectures (e.g., "How to open bananas") which they present in 10 minutes and then examine students in the following 5 minutes. Participants attend the lectures and are graded.

Whenever the participants passed an exam, they are allowed to lecture on the topic (but they are not allowed to stage exams and therefore to train new lecturers). In the end of the academy, all participants are examined from all lectures they attended. In the elaborated closing ceremony participants receive academic or scientific titles. This whole activity develops the *learning* and *presentation skills* and *trains the concentration*.

---

[4] The middle element out of sorted sequence.

Board games. The central component of the game is the board, where one or several team members move the team tokens. The moves on the board or the commodities needed for boosting of the tokens are earned by physical/mental side games. The team has to *plan, find strategies* and *adjust to the individual talents* of its members in order to maximize the gain.

Fighting game. The fighting game is a voluntary game that might be scheduled at the very end of the camp program. It takes place outdoors and its main idea is a battle between the participants and the organizers. It is physically demanding but not as harsh as its name might suggest. A good example of such a game is the so-called sock game. Here the objective is to strip the players from the other group of their socks. The last player wearing his/hers socks wins the game for his/her group. The aim of the game is to *relieve tension* and to *dissolve the distinction between the organizers and participants.*

Graph game. The central goal in this game is to be the first to acquire a certain item. This is not a straightforward process, but rather there are several criteria that need to be met first. Some of the criteria might be dependent on other sub-criteria. The criteria dependencies form a graph[5], which has to be identified by the participants. The standard strategy that the teams come up with is to draw the directed graph and divide its subtrees among the team members. The game promotes abilities such as *cooperation, resource management* and *communication.*

Hiking trip. Depending on the setting of the camp, a hiking trip can be scheduled in the program. During the hike, participants and organizers are getting to know each other, they chat in friendly atmosphere about topics not necessarily connected with computing and thus *strengthening their relations.*

Investigating game. The game is based on solving a Zebra-like puzzle [3] or Smullyan-like logic problem [6]. The outcome of such solution is a password or a meeting place that is needed to proceed in the camp storyline. In addition to the puzzle or the logic problem, a list of extra, quite often funny activities is given. By completing these activities, teams get puzzle clues. The game supports the forming of the *logical thinking* of the participants. Team members have to *make a tactic* when to make an attempt at the puzzle/problem directly and when to use the time and the team-members to obtain a clue.

Night game. This game is not announced in advance, which throws in the element of surprise and thrill. The participants are awakened in the middle of the night, usually because an unexpected event occurred in the camp storyline. They might be asked to rescue a storyline character, follow clues or a map or play an outdoor game to resolve the unfavorable situation. The important rule is not to force the participants to participate. Apart from the thrill of the game, this activity also teaches the participants to *work under pressure and in unforeseen situations.*

---

[5] Usually a directed acyclic graph.

Quiz. A typical quiz team game on various topics, which have mostly nothing to do with CS or mathematics. This is an example of purely relaxing activity, which *relieves tension* and allows for *socialization*.

Role playing game. In the role-playing game [2], the participants impersonate roles in a fictional environment in order to achieve certain goals. Every team plays a separate game session with an organizer that is the game master, i.e. he directs the storyline, provides reactions to the player actions and accounts for all the non-player characters in the game. Participants often try to challenge the game master by creating unexpected situations in which game master needs to improvise to a great length. The game is usually scheduled after the dinner since it is a rest indoor program. Role-playing trains the *communication skills*, *boosts team working* and *encourages the introvert team members* to present their ideas.

Theatre. Theatre is a an evening activity in which the teams prepare a short performance. The performance is often restricted by defined title, topic, artistic style, or it must contain certain phrases or sentences. The aim of the theater is to *relax* and *enjoy the atmosphere through creative brain-storming*. It is therefore rarely counted into the team ranking.

Trading game. The game is usually played outdoors with several outposts approximately 100 to 200 m apart. The goal of the game is to accumulate the wealth which is done by trading different (usually very funny) commodities. During the game the trading outposts change prizes of commodities regularly. The standard strategy that is sooner or later employed by the teams is based on acquiring the commodity prizes at the moment, finding out the formulas to determine how the income can be maximized and employ these until the prizes undergo change. This trains the *strategical thinking* and *time management*.

6   Popularity of Activities

The most important concern during the camp is that the participants associate the fun with CS and learning. To know whether the participants liked the games prepared, we ask them to fill in a questionnaire. This also helps us to prepare subsequent camps. In Figure 2 we present the questionnaire data for the last nine years of camps that we were able to evaluate.

The participants award the activities 0 to 10 points, 10 being the highest mark. For each game in a camp, we use the average of the rating in the visualized graph[6]. The figure for each activity contains the median, quartiles, minimum and maximum of the average, taken from all 18 camps. The number in the brackets represents the number of camps in which the activity was graded.

The games based on physical activity obtained generally lower grades. In particular sports and the hiking trip might be rather unpopular due to their nature of outside exhaustive activities. This would not be surprising since the

---

[6] The median value, that would be more adequate, was not preserved from all the camps.

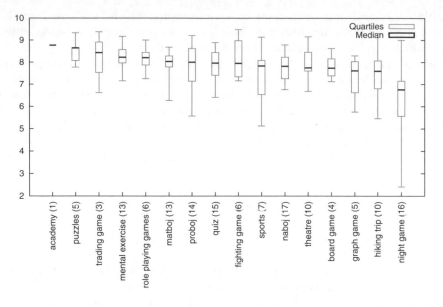

Fig. 2. Popularity of activities in camps between years 1999 and 2008

participants often tend to staying inside doing non-physical activities. Another game of this type is the night game. The game often obtains very mixed grading. This is caused for most of the part by the nature of the game, which many participants might not find attractive. On the other hand, the fighting game is more popular. This is probably due to it being rather short and due to the opportunity to challenge the organizers.

Activities that are related to mental aspects seems to gain a bit higher marks. Their popularity is likely a result of the participants being chosen on the criterion of good ranking in a mental activity of related sort, i.e. KSP.

Academy was arranged only once since its realization is time-consuming, but was received very positively. This is not surprising, since an imitation of academic environment is interesting for the prospective students. Moreover, the game is not physically or mentally demanding and participants can relax while enjoying (or paying no attention to) the lectures.

Since we did not record all data for board games, trading games, graph games, academy and puzzles, these are mentioned in the figure only for the informative purpose.

7   Conclusion

This paper provides a general overview of how a successful CS camp for young people can be prepared. While offering the opportunity for the young people interested in CS to meet and encouraging their learning, the camps also build a number of valuable competences. In the first place, there is the specialized CS knowledge, which forms the core of the educational part of the camp program. Other competences that are trained within the camp program involve various team-building, team-work and social skills, such as time and task management and cooperation between the teams. Last but not least, there are other elements of social nature that the camp program incorporates, e.g., creativity, sports, communication. If well-balanced, the combination of the impulses provided through the camps can successfully help to insure CS community growth.

Acknowledgment

We would like to thank our friends for valuable comments in preparation of this paper. We are grateful to anonymous reviewers for their feedback. We are also greatly indebted to all the generations of organizers that contributed to KSP and camps and who helped to make them what they are today.

References

1. International Olympiad in Informatics, `http://ioinformatics.org/`
2. Role playing game. Wikipedia®,
   `http://en.wikipedia.org/wiki/Role-playing_game`
3. Zebra puzzle. Wikipedia®, `http://en.wikipedia.org/wiki/Zebra_puzzle`
4. Dagiene, V.: The bebras contest on informatics and computer literacy – students' drive to science education. In: Joint Open and Working IFIP Conference. ICT and Learning for the Net Generation, pp. 214–223. Kuala Lumpur (2008)
5. Forišek, M., Winczer, M.: Non-formal Activities as Scaffolding to Informatics Achievement. In: Dagiene, V., Mittermeir, R. (eds.) Information Technologies at School, pp. 529–534 (2006)
6. Smullyan, R.M.: What is the name of this book? Prentice Hall, Englewood Cliffs (1978)
7. Verhoeff, T.: The role of competitions in education, Future World: Educating for the 21st Century (1997)
8. Vygotsky, L.S.: Mind in society: The development of psychological processes. Harvard University Press (1978)

Mission to Mars – A Study on Naming and Referring

Michael Weigend

Westfälische Wilhelms-Universität Münster, Fliednerstr. 21, 48149 Münster, Germany
michael.weigend@uni-muenster.de

Abstract. Algorithms processing complex structures like graphs contain references to entities, which are part of the structure. This contribution first presents an analysis of naming techniques and discusses implications to working memory load. Then some findings of a study on naming and referring in naïve algorithms (involving more than 200 high school students) are presented. The study is based on workshops, in which the students first interpreted a model algorithm and then wrote algorithms by themselves, adopting – to some extent – techniques from the model.

Keywords: Naming, programming, working memory, algorithm, name, intuitive models.

1 Introduction

Referring to entities is an important issue as well in computer programming as in everyday communication. Children start to at things they want to have at the age of one year. A year later they are able to use words like "that" for indicating interesting entities in their environment [1]. People refer to an entity by using a name, which is a symbolical representation. Names have two purposes: identification (making something distinguishable from other things) and addressing (make it accessible).

1.1 Types of Names

A name may be explicit or implicit. Explicit names are verbal expressions that can be said and written down (Mom, E, `sqrt(2)`, `person_1`, `passenger_list[0]`). Obviously all names in a computer program are explicit. Implicit names are gesture-like. They can only be understood taking the context into account. Imagine a customer pointing with her finger to some pastry in a bakery. In this context the gesture is a name for an item she wants to buy. A pin on a map might indicate a certain place. Implicit names may be an integral part of an intuitive model [2]. Imagine moving a finger on a map along roads from one city to another city. This is an intuitive model for processing a graph, searching a path along edges from one vertex to another. It is intuitive, because it is so simple and can be handled in working memory. A program development may start with such intuition. During the implementation the programmer has to realize that the finger serves as a mechanism to address vertices and has to introduce an explicit name for referring to a vertex.

J. Hromkovič, R. Královič, and J. Vahrenhold (Eds.): ISSEP 2010, LNCS 5941, pp. 182–193, 2010.

Explicit (verbal) names can be direct or indirect. A direct name is a name in the most usual sense. It is just a word or a letter, which is connected to an entity in some way (X, Tom, square_a). A direct name of an entity can – for instance – be visualized by a sticky note (lettered with the name) attached to the entity. A direct name is monolithic and completely independent from other entities.

In contrast, an indirect name is constructed using other names. The phrase "my brother's eldest daughter" can be considered as an indirect name, identifying a person (without introducing a new word). In the context of computer programs formal constructions based on arrays, dictionaries or aggregate objects are indirect names. For example the term `passenger_list[0]` is a name for the first item of the sequence `passenger_list` (Python). Function calls can be considered as (indirect) names for the object they return. The expression `sqrt(2)` is used to name the irrational number 1.4.1..., which is impossible to explicate immediately by writing down a floating point literal. Obviously one can construct really complex names. But those are difficult to understand and they are not really useful for problem solving. I am coming back to this in one of the next sections.

1.2 Naming and Typing

Naming is closely connected to typing. In everyday language, when referring to a specific entity we sometimes add a category (type) or consider it as part of the name ("the yellow bird", "Mister Jones", "sphere A"). These examples illustrate that mentioning the type adds meaning to an otherwise meaningless entity name.

Programming languages like Java require declaring a variable before using it in a statement by allocating a type to the name. Dynamic languages like Python adopt "Duck Typing" and do not require explicit declarations of variables. Still, they connect types to names. When assigning an object to a name as in

```
x = 1.0
```

the system deduces the type of the object (`1.0`) and attaches the type (`float`) to the name (`x`). Otherwise expressions like `x + y` could not be interpreted. The type may be deduced from the lexical structure of a literal. In Python all literals consisting of decimal digits and exactly one point somewhere in the middle represent an object of the type `float` (floating point numbers).

1.3 Functions – Naming Activity

To define a function (or method) means to attach a name to some activity. A function must represent an intuitive, Gestalt-like piece of knowledge to be useful for solving an algorithmic problem, Because of the limited working memory (see below) there is no cognitive advantage to use a function call, when you have to memorize (additionally to the name) how this function works. I will come back to this later.

1.4 Naming and Working Memory

Both reading and creating algorithms are quite demanding cognitive activities. Cognitive psychology assumes that working memory is a central component for this kind of information processing [3], [4], [5]. Working memory is a mechanism to get very quick access to mentally stored information but its capacity is very limited. Names are basically verbal information. According to Baddeley's model, working memory contains a subcomponent for storing verbal information, called the phonological loop [3], [4]. It is analogous to an audio tape recorder loop that is able to store verbal information, which can be articulated in not more than two seconds. This implies that just one complex reference like "my brother's eldest daughter's dog" might block the working memory completely and prevent any further information processing. On the other hand short verbal identifiers like in "a square plus b square equals c square" are of advantage.

In imperative computer programming (using languages like C, Java, Python, Pascal) assignments can be used to name entities. In a statement like

```
letter = flight.passengers[i].name[0]
```

a complex reference (indirect name) is replaced by a simple direct name. Such assignment facilitates programming as well as program comprehension, because working memory load is reduced. The assignment empowers the programmer to use just the word *letter* instead of a complex reference to develop an algorithm, in which this referred entity is processed. During the problem solving process it is now possible to keep other relevant chunks of information in working memory.

Using names in a smart way is essential in software development. During software design [6] or software refactoring (in Extreme Programming [7]) much effort is put into developing a consistent framework of names. Naming appears to be an important part of coping with complexity. But it is more than just a programming skill. When solving a task in science education, mathematical modeling starts with identifying and naming relevant entities ("Let c_0 be the concentration of the hydrochloric acid in the beaker..."). Additionally science provides some kind of international style guide for naming. For example in physics, the letter F - or some string starting with F, like $F1$ or $F(car)$ - is supposed to refer to a force.

Smart naming seems to be not just a technical skill of specialists but deeply embedded in general competences like problem solving and communication. In the face of this the question arises to what extent students without former informatics education are able to use naming techniques, when they have to write algorithms. What is easy to them? What are the difficulties? Knowledge in this field seems to me quite interesting for computer science teachers, who are designing lesson plans and classroom activities for introducing programming concepts.

2 Design of the Study "Mission to Mars"

In summer 2009 I conducted workshops in two German comprehensive schools with altogether more than 200 students from grades 9 and 10 at the end of the term. The workshops were called "Mission to Mars", since they were embedded in a little story related to our red neighbor planet. The first two "missions" were about instructing a robot to build a habitat on Mars. In the third "mission" the robot had to exchange certain parts of a factory or power plant.

Each 45-minutes-workshop consisted of four phases:

Phase 1: The students got an algorithm written in everyday language. They had to read it and execute it using pencil and paper. This algorithm contained a variety of different ways to refer to entities within a geometric structure. I did not encourage the students to talk to their neighbors during phase 1. Nevertheless, they were explicitly allowed to communicate, when they had problems. Some did so. Thus, in some cases the results of phase 1 do not represent individual achievement but the collective competence of a group of students sitting together.

Phase 2: The students got a task and had to write an algorithm on their own. This algorithm was similar to the one they got in phase 1. There were eight different versions of the task and they were distributed in a way that within a group of students sitting together at one table each task was unique. Again the students were allowed to talk. In some cases they definitely stole some ideas from their neighbors. Nevertheless, it was impossible just to copy a solution, since all neighbors had different tasks.

Phase 3: Each student exchanged his or her algorithm with someone who had got a different task and who were (ideally) sitting at a different table in the class room. Each student tried to execute the algorithm she or he had got from a classmate. During this phase no further communication between these two persons was allowed.

Phase 4: The pairs, who had exchanged algorithms, came together and discussed the results. Are there differences between the produced result (the interpretation of the algorithm) and the intended result? What are the reasons for eventual differences?

At the beginning of the workshop the participants were informed about this structure. This was important, because everybody should have the feeling that the written text must be understood by a classmate. There is some empirical evidence that assumptions about the addressee's knowledge have influence on the accuracy of verbal messages [8].

The results of phase 1 and 2 were collected and evaluated later in order to get answers to two general questions: (1) Did the students understand the naming and referring techniques used in the given algorithm of phase 1 (model)? (2) Which of the techniques in the model did they adapt and adopt when writing their own algorithms?

3 Mission 1 – Naming and Referring

105 students participated on workshops entitled "Mission to Mars 1" in summer 2009 (grade 9, average age 15.3 years, 57 male, 47 female, one person did not tell the gender). In phase 1 the students had to interpret instructions for drawing the ground plot of a fictive Mars habitat. The algorithm consisted of two procedures. The second procedure was called "to mark a star" and explained how to allocate natural numbers to the tips of a star with an arbitrary number of beams. The first procedure explained how to draw the habitat. Here are a few instructions:

(1)   Draw a five point star. One of the tips points upwards. Write the letter A in the middle of the star.

(2)   Draw a four point star with four beams a little bit below star A. One tip of the new star points upwards. Write the letter B at the star.

(3)   Mark star A. Mark star B.
(4)   Draw circles around the tips number 1, 2 and 5 of star A.
(5)   Draw circles around the top and the right hand tip of B.
(6)   Put your finger on the circle, which is at tip number 5 of A. Draw a second circle left to this circle.
(7)   Write the word Sauerstoff (oxygen) next to the right hand side of the circle, which is at tip number 2 of A.

Figure 1 (left picture) shows a correct interpretation of the model algorithm.

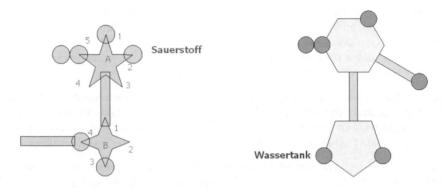

Fig. 1. Left picture: One of several possible correct interpretations of the algorithm of phase 1. Right picture: One of the eight different tasks in phase 2.

The model algorithm contained different types of naming and referring to entities. The first two instructions attached explicit direct names A and B to stars. Later these names were used to address the stars. In instructions 6 the interpreter is asked to put a finger on a certain entity. This gesture is an indirect name. It is used to divide a complex instruction is into two simpler parts: 1) Find a certain entity. 2) Draw something next to it.

An easy way to refer to a certain tip of a star is to use numbers (e.g. instruction 4). A drawback is that you have to attach numbers to the tips first. That means you model the star as an array of tips. Alternatively one can use geometrical attributes like "left", "top", "bottom" like in instruction 5.

62% of all students were able to execute the algorithm without any mistake (boys: 61%, girls: 62%). Another 23 % made not more than one mistake. This indicates a pretty good comprehension of the instructions.

In phase 2 the participants got a ground plot (like in figure 1 right hand side) and had to write a verbal algorithm explaining how to draw it. Instead of stars the plots contained polygons. There were eight different versions with exactly the same complexity. To what extent did the students adopt the reference techniques from phase 1?

32% of the students used explicit direct names to identify components of the picture. Most of them (87%) used exactly the same names (A, B) as in the model algorithm. Names were introduced using phrases like "Call this pentagon A" (13%) or through write commands like "Write at the pentagon the letter A" (20%). 9% of the students used a phrase like "Put your finger at ..." (indirect name) to split a complex reference into two simpler parts.

66% of the students adopted the idea of indexing the corners of a polygon and referred to specific corners using their numbers. But before using indices you have to define them. Indeed 90% of the students applying this technique wrote some instructions for numbering corners. Most of them (55%) did this in a a complete and correct way. Just 25% of all students using numbers defined a separate procedure (with a header like "to mark a polygon") for attaching numbers to corners. It should be mentioned that there were no significant differences between boys and girls.

4   Mission 2 – Using Names for Activities

52 students participated in Mission 2 (grade 10, average age 16.1 years, 32 boys, 19 girls, one person did not tell the gender). The model algorithm in phase 1 consisted of four procedures with one to six instructions each:

To draw a Mars habitat
To draw a living unit
To draw a laboratory
To draw a tank

Like in Mission 1 the instructions included direct and indirect names for entities, but the predominant feature was the extensive use of procedures. A procedure is a function that does not return an object (resp. returns just an empty object). The main algorithm "To draw a Mars habitat" contained procedure calls, which corresponded to some geometric figure (like the plot of a "living unit" of a Mars habitat) each. It started like this:

(1)   Draw a living unit in the middle of your paper. Call it W1.
(2)   Draw a second living unit below W1. This new living unit is called W2.
(3)   ...

Most students had no problems to interpret the instructions. 75% of them draw a plot, which was completely correct and another 12 % made just one mistake. On the other hand, 10% of the solutions contained more than 3 errors.

In phase 2 each student got one of eight different plots like in figure 2. Each plot contained three instances of an aggregate (consisting of four simple shapes). The pictures were colored in a special way to highlight the aggregates.

A structured algorithm (like the model algorithm in phase 1) would define a procedure for drawing an aggregate and call this procedure three times. But only 2 out of the 52 students defined procedures in their algorithms and thereby attached meaningful names to coherent bunches of activities.

Most students (73%) numbered all instructions (like in the model algorithm). The number of an instruction can be considered as a name, which identifies an instruction. 6 students referred to an instruction using its number (e.g. "Execute step 3 another time").

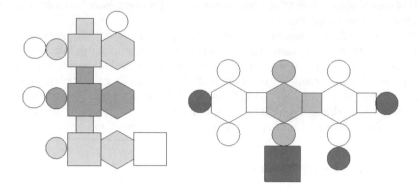

Fig. 2. Two examples of plots from phase 2 in "Mission to Mars 2"

29% of the students used subheadings (like "First part of the habitat") thus separating sections of the algorithm. A subheading can be seen a name for the following block of instructions. But this name is never used to *refer* to the activity like in a procedure call. Inserting subheadings is a way to structure a program text on a lower level of abstraction. There is no distinction between defining an operation and executing it.

Again – out of 52 students, just two individuals (less than 4%) wrote a structured algorithm consisting of at least two procedures. Although all participants had interpreted and elaborated a model in phase 1, which illustrated how to do it, only two adopted this technique. Why? Before presenting a possible explanation let me discuss what makes it difficult to *interpret* a structured algorithm containing function calls.

Consider a person – let us call her Sandra – interpreting an algorithm containing function calls. You can distinguish two cases. First case: Sandra is familiar with all functions and can execute them without conscious mental effort. The central executive of the working memory benefits from this automaticity (fluency) and is allowed to attend to higher level processes related to the algorithm [5, p. 26].

What if Sandra is not familiar with the concept of a called function (second case)? She might rehearse it first, until she reaches some level of automaticity and the function name has got a meaning to her. In fact during phase 1 of "Mission to Mars 2" some students first practiced the involved procedures (by drawing plots of a "laboratory" and a "living unit") before they started to execute the main algorithm.

If Sandra does not manage to keep the meaning of a function in her working memory she has to use the written definition as an external aid. This is a disadvantageous split-attention situation [9], causing an increased cognitive load: Algorithmic information, which belongs together, is presented at different places and has to be mentally combined by the reader.

From algorithm interpretation back to algorithm creation: Structuring an algorithm by introducing functions is a complex and difficult task, when the goal is not just to divide a long text into smaller pieces but to increase the understandability. It implies chunking activity to *meaningful units*, which are easy to grasp. This is of help not only for the reader of a program text but also for the developer, who has to guarantee the logical correctness.

A function is a coherent, meaningful activity with a name. Finding useful functions (which increase the understandability) means additional development effort and thus is a barrier slowing down development speed. Most of the students participating in the "Mission to Mars 2" apparently avoided this additional effort.

However, structuring system activity by defining functions (or classes with methods) is essential, when you have to develop large software systems. Otherwise programmers would not be able to cope with the complexity. In big projects functions and class structures are necessary even at the cost of understandability, since they better other quality properties like adaptability (changeability) and testability [6].

5 Mission 3: Referring to Entities within a Complex Geometry

Mission 3 focused on reference strategies. In phase 1 the students got a picture of a fictive power plant on Mars and instructions referring to ten parts of its three-dimensional structure (see table 1). They had to identify entities by writing numbers on the picture. According to the cover story these parts got malfunctions and had to be replaced by a robot.

Table 1. Instructions referring to entities within a three-dimensional structure and number of students, who were able to interpret them. Results from workshops with 49 students from German comprehensive schools (grade 9, average age 15.5 years, 25 boys, 21 girls, three persons did not tell their gender)).

Nr.	Instruction	Correct solutions
1	The pyramid in the right corner of the platform is called corner pyramid. Write number 1 on the corner pyramid.	49 (100%)
2	There is small light sphere on top of a bigger dark sphere. Write number 2 on this sphere.	47 (96%)
3	A Threesphe (German: Dreiku) consists of three spheres, which are connected vertically. One of the Threesphes has got a dark sphere in the middle. Write number 3 on the cube, which is beside this Threesphe on the platform.	47 (96%)
5	A Cylsphe (German: Zylku) is a cylinder with a sphere on top. Write a 5 on the cube, which is on top of the Cylsphe, which is in front of a pyramid.	45 (92%)
6	When you move from the corner pyramid along the edge of the platform to the left, you reach a cube. Write 6 on this cube.	45 (92%)
7	X is the cube at the Threesphe, which is beside a pyramid. Write number 7 on X.	41 (84%)
8	Put your finger on the big sphere in the middle of the platform. At this sphere there is an arrow pointing to a pyramid. Write an 8 on this pyramid.	49 (100%)
9	A Fourcyl is a flat cylinder with four Cylsphes on it. Look for the sphere, which is floating over the Fourcyl. Write number 9 on this sphere.	47 (96%)

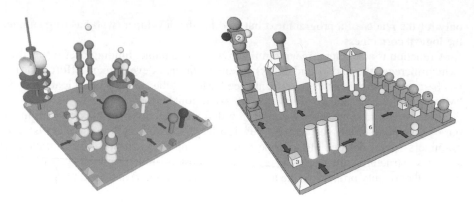

Fig. 3. 3D-structures used in phase 1 (left picture) and phase 2 (right picture) in "Mission to Mars 3"

The instructions contain some ideas how to refer to entities within the structure:

(1)   Explicit direct names (*A*, *X*, *corner pyramid*) are introduced referring to certain entities within the structure. These names especially facilitate path-like references. For instance the name *corner pyramid* is used several times as starting point for the description of paths to different entities. Additionally longs paths with many vertices can be split in smaller parts by using a name for an entity in the middle.

(2)   Some instructions contain the phrase "put your finger on …" thus using an implicit name for an entity. This technique should help to keep track of a path because it reduces working memory load. When you have put your finger on a thing you need not to remember it.

(3)   Type names are introduced to name aggregates consisting of simpler geometrical bodies ("A Cylsphe is a cylinder with a sphere on top") including a three-level structure in instruction 9 ("A Fourcyl is a flat cylinder with four Cylsphes on it."). Types are useful for aggregate-oriented references (see below) which start with specifying an aggregate and then refer to parts of it.

In average the students interpreted 95% of the instructions correctly. The (apparently) most difficult task (7) was solved correctly by 84% (see table 2).

5.1   Using Names for Entities and Types

In phase 2 the students got a picture showing a completely different structure, representing a fictive factory on Mars (see figure 3 right picture). Six parts got numbers from 1 to 6. The students had to write instructions to identify these parts. In phase 3 – again – a classmate had to read and execute these instructions using the same picture, but without numbers.

Although the students apparently had no problems to understand the instructions of phase 1, only a minority of made use of explicit names themselves. 23 out of 49 students used some kind of naming concept (including implicit naming by pointing with a finger) for individual entities. But just three of them introduced an explicit name, which they used in several instructions. Another five students used the number of an entity they already had identified to indicate the starting point of a path to a different entity ("Put your finger on part number 3 …"). 12 participants used the name corner pyramid from phase 1 without explaining it. In the picture of phase 2 there was just one pyramid situated in a corner of the platform, so it was quite clear which object was meant. 13 students wrote instructions containing phrases like "put your finger on …", thus using an implicit name concept.

Only three students invented and explained new names for types of aggregated bodies in the style of phase 1. Examples are "Würzel" for a figure composed of a cube (German: Würfel) and a Cylinder (German: Zylinder) and "Würgel" for an aggregate of a cube and a sphere (German: Kugel). Another six students just used a new name without defining it properly. (In some cases it can be assumed that they stole the idea from a neighbor.) Three students used a type from phase 1 (Cylsphe).

5.2 Referring Strategies

I distinguish three different strategies to refer to an entity within a structure: Attribute oriented references, path oriented references and aggregate oriented references.

(1) An attribute-oriented reference describes an entity as an instance of a type and then specifies some attributes like size, color and geometrical information. It does not depend on other entities and is monolithic in this regard. Consider this example from phase 1: "There is small white sphere on top of a bigger dark sphere." A person interpreting this instruction has to keep all this information in his or her working memory while searching for a matching entity. This makes it difficult to interpret it.

In computer programming this strategy is used in iterations over collections (sequences, sets) searching for a specific element by checking certain properties.

A problem is that the author of such reference must be sure that it leads to only one unique entity within the regarded scenery. Thus in some cases a attribute oriented reference might cover more attributes than necessary. The following example was written by a 15 year old boy: "The cube is glued at a cylinder, which is on top of a bigger grey cube, which stands upon four white cylinders. Additionally there are two white pyramids on top of the big cube." 27 students (56%) used attribute-oriented references in phase 2.

(2) A path-oriented reference starts with some entity, which is easy to identify. It uses an existing direct name (like corner pyramid) or an attribute-oriented description to identify this start entity. Then it describes a relation to another entity and so on until finally the goal entity is reached. Four instructions of phase 1 (including 4, 6, 8) were path printed. 43 students (90%) used references of this kind and 28 (58%) even described paths with three vertices or more. This is an example written by a 15 years old boy: "The arrow points to a small cube. This is in front of a bigger cube. On the bigger cube there is a sphere. Write number 5 on this sphere." Usually the relationship between two entities on the path is geometric in nature. Phrases like "in front of" or "on top of" are used. Navigation along a path is associated with movement from one place to another. Some instructions of phase 1 supported this notion by using phrases like "move to …" or "move your finger to …" 30 students (61%) used move-phrases in their instructions.

In many cases the usage of a path means a reduction of working memory load. This is because the path starts with an easy identifiable entity and the movement from one to the next entity is easy to describe as well. When you have reached an entity during navigation you can point your finger on it and drop all the information chunks from your working memory, which you needed to get there. Thus, interpreting and creating a lengthy path might require less working memory load than interpreting or creating a more concise attribute oriented reference. Using a finger for navigation means adopting implicit naming. Some path-oriented references in the model algorithm (phase 1) made use of explicit names for entities on the path like the letters A and B in instruction 4 or the name *corner pyramid* to indicate a starting point. But *not a single student* used an explicit name for an intermediate vertex on a reference path. It should be mentioned that in phase 1 six participants wrote letters (A, B or X) at the corresponding entities on the picture of the "power plant". In other words: These students used explicit names of the instruction to construct external aids that reduce working memory load.

(3) An aggregate-oriented reference starts with the description of a structure composed of two or more simple bodies. Then it refers to a part of this aggregate and so on until the target has been specified. This is an example, written by a 15 years old boy: "You see a row of cubes. On top of each cube there is a sphere. Beside one cube with a sphere on top there is another sphere. Write number 6 on the sphere, which is on top of this cube."

The model algorithm in phase 1 made use of aggregation by defining and naming some types of composed structures. But phase 1 did not contain a clear example of this referring strategy.

An aggregate-oriented reference is similar to a path. In fact it could be considered as a path from the outside (aggregate) to some place within. It assumes a tree-like structure of the 3D-geometry. The whole scenery is the root and simple bodies (cubes etc.) are the leaves. Aggregation is an important concept in Object Oriented Modeling (OOM). In contrast to path-oriented references, an aggregate-oriented reference requires a structural analysis of the scenery – similar to OOM. So it is quite surprising that 28 students (57%) used at least one aggregate-oriented reference in phase 2.

6 Conclusion

In this final section I am going to sum up the empirical findings and then add a few thoughts about CS teaching, which are inspired by the observations.

First of all the study suggests that there is an immense discrepancy between algorithm comprehension and fluency in algorithm creation. It is not enough just to understand an algorithmic concept to be able to apply it.

Creating explicit and direct names for entities, types and activities seem to be something difficult and unusual for naïve algorithm writers. Although the 9th- and 10th-graders observed in this study could understand and interpret algorithms with explicit names pretty well, most of them avoided introducing names, when it was about to create algorithms. Using entity names to break up complex references and to make explanations simpler, seems to be something unusual for them. Computer programming relies on explicit naming. The *idea* of an algorithm might be better understood, if explicitness is avoided. Explicit names sometimes destroy the intuition, because they make the model too complex [2]. But in order to explicate the idea and develop a program or an algorithm that is understandable to other humans, elaborated explicit naming is essential. The study showed that some aspects of naming (especially naming activity by defining functions) are almost completely unknown to students of the observed age group. These things apparently are not learned in general education outside CS classes. Thus teaching formal naming concepts in computer science lessons might lead young people to a new dimension of verbal expressiveness.

Students understand a language concept in a much deeper way, when they have experienced the explication and communication problems, which are solved by this concept. In this regard the classroom activities presented in this paper might be useful for teaching. In a real lesson there should be a debriefing after phase 4, in which language problems, which came to daylight in the earlier phases, are discussed. Experiences and insights gained in such "communication experiments" may serve as anchors for introducing programming concepts.

References

1. Matthews, D., Lieven, E., Tomasello, M.: How Toddlers and Preschoolers Learn to Uniquely Identify Referents for Others: A Training Study. Child Development 78(6), 1744–1759 (2007)
2. Weigend, M.: Intuitive Modelle der Informatik [Intuitive Models in Computer Science], Universitätsverlag Potsdam (2007)
3. Baddeley, A.: Recent Developments in Working Memory. Current Opinion in Neurobiology 8(2), 234–238 (1998)
4. Baddeley, A.: Working Memory: Looking Back and Looking Forward. Nature Reviews Neuroscience 4, 829–839 (2003)
5. Dehn, M.J.: Working Memory and Academic Learning. John Wiley & Sons, Hoboken (2008)
6. Sommerville, I.: Software Engineering, 5th edn. Addison-Wesley, Harlow (1997)
7. Beck, K.: Extreme Programming Explained. Addison-Wesley, Boston (1999)
8. Krauss, R.M., Fussell, S.R.: Perspective-taking in communication: Representations of others' knowledge in reference. Social Cognition 9, 2–24 (1991)
9. Ayres, P., Sweller, J.: The Split-Attention Principle in Multimedia Learning. In: Mayer, R.E. (ed.) The Cambridge Handbook of Multimedia Learning, pp. 135–146. Cambridge University Press, Cambridge (2005)

Long-Term Development of Software Projects – Students' Self-appreciation and Expectations

Cecile Yehezkel[1] and Bruria Haberman[2]

[1] Davidson Institute of Science Education, The Weizmann Institute of Science,
Rehovot 76100, Israel
cecile.yehezkel@weizmann.ac.il
[2] Computer Science Dept., Holon Institute of Technology, and
Davidson Institute of Science Education, The Weizmann Institute of Science,
Rehovot 76100, Israel
bruria.haberman@weizmann.ac.il

Abstract. The "Computer Science, Academia and Industry" extracurricular program has been operating at the Davidson Institute of Science Education for the past few years. The program, designed especially for high-school students majoring in computer science (CS), aims to bridge the gap between schools and the "real world" of computing and provides students with the opportunity to meet with leading computing experts from academia and industry. The program includes a preliminary stage of enrichment meetings and an advanced stage in which students develop software projects. A long-term formative evaluation of the program has been conducted regarding students' attitudes towards the "different-from-school" style of learning, and their performance in developing projects. This paper describes a specific study aimed at determining how students' self-appreciation and their initial expectations affect their readiness and aspiration to complete a long-term comprehensive project.

1 Introduction

During the last two decades, two elective programs, in computer science [5] and software engineering [2], have been operating in Israel as part of the formal secondary education. The aim of these programs is to expose young students to the fundamentals of computing, and to motivate them to seek expertise in this field. Even though the programs have evolved over the years, a gap still exists between the school's educational milieu and the "real world" of computing regarding content, learning style, and the professional norms typical of software development processes [16].

The basic fundamentals and core technologies that are introduced in school constitute the basis for students' understanding of the field of computing; however, they rarely include state-of-the-art computing research and development as well as new, rapidly evolving directions in the field.

The traditional style of teaching and learning in school is usually designed so that students can acquire explicit knowledge based on a thorough understanding of the topic learned. However, this approach alone might fail to teach students to become self-learners who are capable of reliably navigating in the world of rapidly growing

J. Hromkovič, R. Královič, and J. Vahrenhold (Eds.): ISSEP 2010, LNCS 5941, pp. 194–205, 2010.

knowledge [11,12]. School projects enable students to experience software design and development [5,6,10]; however, they do not resemble actual software engineering industrial processes, and the products are rarely applicable to real-world situations [14,15,16].

The above considerations motivated us to develop the "Computer Science, academia, and industry" extracurricular program, designed especially for talented high-school students who major in computer science. The program aims at exposing students "directly by leading experts" to state-of-the-art research, advanced technologies, software engineering methodologies, and professional norms. The program is planned so that students will experience a style of learning that will prepare them to successfully navigate in the ever-changing, dynamic world of digital and other rapidly increasing knowledge.

The two-year program blends formal and informal learning and includes enrichment meetings, field trips (Stage A, the first year) and software development projects supervised by experts (Stage B, the second year). Through these activities, the program aims at bridging the gap between computing and software engineering education and the fundamentals of computer science taught at the high-school level. A detailed description of the program (related to its setting and contents) is presented in [16]. The program has been conducted for the last five years and serves as an example of a successful partnership among academia, industry and the Ministry of Education. Eight-hundred and ten students participated in enrichment activities; 138 students developed high-level software projects. Recently a new (6th) cycle of the program was started.

A long-term formative evaluation of the program has been conducted regarding: (1) students' attitudes towards the "different-from-school" style of learning that characterizes the program, and (2) their performance in developing projects. So far, the study's findings have indicated that the program contributes to developing a culture of learning befitting the dynamic "real" world of computing, thus providing the students with an entry point into the computing community of practice [7,8,16]. In this paper we focused on a specific study aimed at determining how students' self-appreciation and their initial expectations affected their success in completing their project (initial findings of this study are presented in [9]).

2   The Project Development Activity

According to computer science educators, the role of projects in the curriculum is of major importance, since it is a means of promoting effective learning. Project-based learning enables students to construct knowledge and to enhance cognitive and reflective skills. It enables students to encounter real-life experience as a project developer and it also encourages them to become creative and independent learners [1,3]. Project development enables students to demonstrate their mastery of skills appropriate to professional practice [4,6,10].

During the second year of the "CS, academia & industry" program, the students develop comprehensive software projects in a variety of scientific and industrial fields under the apprenticeship-based supervision of professional mentors. Aimed at exposing the students to state-of-the-art computing research and development as part of their preparation for the project development activity, the program offers a wide spectrum of topics, some of which are specifically dedicated to software engineering methodologies and professional norms. In addition, meetings related to time management and role-playing activities are conducted.

The project development process lasts 9-10 months according to a given time table. Some of the students actually participate in "real" industrial/academic projects, thus solving "real-world" problems for a real client; others utilize advanced development tools. During the development process, the students need to acquire theoretical and technical knowledge, such as initial acquaintance with the subject, studying the theoretical background needed for developing the project, searching for relevant classic algorithms, studying a suitable programming language, and selecting an environment appropriate for development. To support the development process, meetings involving the entire Stage-B group of students are devoted to software development issues that experts present and discuss. During these meetings the students are requested to submit written sub-products (e.g., specification documents and a mid-term report) and to discuss their progress with the whole group. Students are also encouraged to ask for help through a technical-support forum run by a counselor.

At the end of this process, the students have to submit a working system and a written report that describes the developed software and the outcomes of each stage of the development process. In addition, the report must include a printout of a documented code. Examiners nominated by the Ministry of Education evaluate the students' final products and the students' final grades are considered as part of their high-school Matriculation Diploma.

2.1 Underlying Principles

The project development activity is designed for those "cream of the crop" students who exhibit the following characteristics: high motivation, creativity, self-learning and inquiry ability, persistence, consistency, and the ability to follow a time table. The selection of students for this activity is based on the following criteria: (a) the teachers' recommendation; (b) the applicant's resume, which should provide information about his CS knowledge, programming experience, knowledge of programming languages, participation in other relevant enrichment programs, and experience in developing software projects; and (c) the applicant's ability to persuade us that he is seriously interested in developing the project, and that he is capable of successfully completing the project development assignment [16].

Our aim is that the activity will foster and enhance students': (a) skills such as creativity, self-learning, curiosity, inquiry ability, innovation, system-level perception; (b) good habits such as time-management ability, communication skills, and the ability to follow specifications; and (c) professional knowledge. We believe that this can be achieved by optimal student-mentor matching as well as good monitoring. The role of the mentor is twofold: (a) to provide the student with guidelines and resources for acquiring theoretical and technical knowledge needed for developing the project; (b) to guide and to control the student's progress in various stages of developing the project, for example, checking whether the product addresses the initial specifications and requirements; checking whether the student progresses according to a planned time table; assessing the use of design methods; and assessing the quality of the programming.

Furthermore, the mentors serve as role models for the students, as mediators of information related to norms as well as to actual problems that members of the computing community of practice encounter.

One main underlying principle of our instructional model is to provide diversity. We believe that it is important to enable students to choose the project's subject from a variety of subjects suggested by the mentors. The possibility of choosing a subject from a wide range of subjects may increase students' intrinsic motivation to develop a project, and encourage them to reach their potential. This goal can be achieved if a diverse group of mentors are recruited (e.g., scientists and engineers from academia and the hi-tech industry, and CS/SE graduate and PhD students). Another aspect of diversity relates to the mentoring method, the student-mentor communication, and the student's style of self-learning. In our program we opted to provide this type of flexibility to facilitate optimal student-mentor interaction and performance. This is in contrast to a "regular school setting" where only one mentor is available for all students, and it is common that the projects of students that belong to the same class resemble too much the projects of their classmates [14].

Another primary principle is that the teachers will be actively involved in supporting the students throughout the entire development process. Actually, the challenge is to create effective cooperation among all four types of participants: (a) the students, (b) the mentors, (c) the academic management team of the enrichment program, and (d) the teachers.

Student-mentor matching is the key for successfully developing a project. To establish optimal matching, we developed the following employment fair model: a special meeting is conducted, where, in a plenary session, the mentors present to the students a variety of project subjects for which they can serve as advisors. After the presentation, a face-to-face mentor-student interaction takes place where students are asked to present to mentors their "CV-like" applicant's resume. The process ends when all possible interactions take place. During the interaction, the students ask the mentors questions about the suggested projects and examine whether the topics seem attractive and can be dealt with and whether they want the mentor to guide them. At the same time, the mentors implicitly investigate whether the students are qualified enough to develop the project that they suggested. Next, the students are required to submit a list of projects in order of their preference, and the mentors are asked to choose students according to their assessment. Finally, the managers of the program perform the mentor-student pair matching [16].

2.2 Implementation

The program has been run for the last six years. Eight-hundred and ten students participated in the preliminary enrichment activities; of these, 333 were selected to participate in the project development activity; 138 students completed their comprehensive software projects. Recently 93 students started to develop projects and are supposed to submit the projects in April 2010.

The subjects of the students' projects are usually related to the topics of the enrichment meetings and actually reflect the mentors' background. Most of the projects mentored by industry representatives have practical characteristics; for example, computerized homes,

Table 1. Participation in the program

Years	# Schools	# Students Stage A*	# Students Stage B **	% Grad. ***
2004-2006	9	71	25	52%
2005-2007	20	140	50	56%
2006-2008	30	180	80	56%
2007-2009	30	210	85	61%
2009-2010	30	220	93	Still working

*   Enrichment meetings
**   Project development activity
***   Students who successfully finished their project

managing a multimedia-shop, programming a robot, missile detection, and computerized aquarium care. On the other hand, the projects mentored by CS faculty and research students focus on theoretical or research-based subjects such as computerized graphics, image processing, automatic text categorization, modeling-based development of a control system, disassembling and reassembling DNA, simulation of the theory of natural selection, utilizing neural networks for the recognition of characters in a picture, and games based on artificial intelligence.

3   The Study

A long-term formative evaluation of the program has been conducted regarding: (1) students' attitudes towards the "different-from-school" style of learning that characterizes the program, and (2) their performance in developing projects. So far, the study's findings have indicated that the program contributes to developing a culture of learning befitting the dynamic world of industrial computing, thus providing the students with an entry point into the computing community of practice. We found that project development experience under the supervision of professional experts may motivate students to acquire more in-depth knowledge in computing, as well as promote creativity, enhance self-learning and inquiry ability, and contribute to establishing professional norms [7,8].

For example, one main goal of the study was related to students' assessment of the resources that they use for developing projects. The findings indicated that during the entire development process the students exhibited self-efficacy, since they relied more on themselves than on other resources. For example, to achieve adequate acquaintance with the needed theoretical knowledge, self-studying and the web were perceived as the most significant resources, which may imply that the mentors' guidance inspired the students' self-inquiry and self-study. However, during the problem-solving activities, students relied more on the mentors than on bibliographic resources [7].

Another goal was to assess students' project management style and their communication with the mentors. We diagnosed 3 profiles of student behavior: (a) students with a large amount of independent work along with a few meetings with their mentors, which started after a period of self-studying and later on continued with a slight increase towards the completion of the project; (b) students who choose to be coached intensively by the mentor during the whole period and apparently were unable to achieve autonomous self-investment; and (c) students who combined self-studying with regular on-going mentoring [8].

In this paper we focused on a specific study aimed at determining how students' self-appreciation and their initial expectations affected their readiness and wish to complete a long-term comprehensive project. The motivation underlying this particular study was to try to understand the reasons for the high number of drop-outs (~55%) from the project development activity. Specifically, we were interested in checking whether the outcome of such a study could: (a) give us some insight into how we could better select students capable of completing their projects; and (b) assist us in improving the mentoring/monitoring model.

3.1 Methodology

The study population consisted of 85 students who were accepted (from a larger group of candidates) to the project development activity in May 2007. All students started to develop a project; however, only part of them completed the process. We divided the students into two groups: (a) those who finished their projects (Graduates; N=45), and (b) those who dropped out at any stage of the process (Drop-outs; N=30).

As part of their application to Stage B, almost all candidates (N=75) filled out a questionnaire to check their preliminary approach to the project development activity. The questionnaire was multipurpose; besides being an integrated component of the students' participation in the program, its collected data helped us perform our study. Importantly, one has to take into consideration that the questionnaire was composed of open questions; hence, all students worded on their answers and expressed "important items" that they had in mind and were not influenced by a pre-determined selection of items (as is common in closed questions). The questionnaire related to the following: (a) Motivation- students' initial motivation to develop a project; (b) Qualification- personal aptitudes considered by the student as valuable for developing a project; (c) Expectations- expected benefits of developing a project; (d) Domain of interest- the domain on which the student would like to focus; and (e) Time investment- pre-assessment of the time needed for developing the project (presented in terms of hours per week).

This kind of open questions has the advantage of freeing the students to elaborate original personal answers; however, analysis of the required answers is more complex and mainly provides insights on trends rather than rigorous statistical results. Accordingly, the character of our research was essentially qualitative because we had to interpret students' answers to the open questions, identifying common items and categorizing them. Students' answers to questions (a), (b), and (c) were verbally analyzed and categorized. Quantification of the qualitative data was performed by identifying recurrent items.

3.2 Findings and Discussion

The identified items of questions (a), (b), and (c) and their frequencies are presented in Table 2 and Figure 1. Examination of Table 2 reveals interesting relationships between items associated with different super-categories (e.g., *Motivation*, *Aptitudes*, and *Expectations*). We found that all items associated with *Expectations* (6 items) were included in the set of items associated with *Motivation*. The students choose to relate these items either to the *Motivation* question or to the *Expectations* question, or to both. Retrospectively, we realized that we should have improved the questionnaire's wording to emphasize the difference between what we meant as intrinsic motivation and extrinsic expectation [13]. As a result, after a preliminary analysis of the data, we decided to "logically" merge items per student associated with both supercategories ({*Motivation* items}∪{ *Expectations* items}).

Table 2. Frequencies of categorized items in students' answers

Question: Why do I want do participate in stage B?		
Motivation	**Graduates**	**Drop-outs**
Matriculation units	**25%**	23%
Creativity	2%	**13%**
Acquiring software development experience	**23%**	10%
Knowledge enrichment	93%	**97%**
Acquaintance with industrial proficiency	**27%**	23%
Acquaintance with professionals (experts)	**20 %**	17%
Fondness of the domain	20%	**27%**
Self-satisfaction	**7%**	3%
Question: Which qualities that I posses will help me to complete the project?		
Aptitudes	**Graduates**	**Drop-outs**
Creativity	**25%**	20 %
Having experience	23%	**27%**
Knowledge	7%	**10%**
Diligence	77%	**90%**
Fondness of the domain	**30%**	13%
Fulfilling requirements	0%*	**13%***
Fulfilling standards	0%*	**10%***
Self-learning	**64%**	60%
Question: What benefits can I expect from the project?		
Expectations	**Graduates**	**Drop-outs**
Matriculation units	25%	**27%**
Acquiring software development experience	**43%***	20%*
Knowledge enrichment	91%	**93%**
Acquaintance with industrial proficiency	16%	**20%**
Acquaintance with professionals (experts)	14%	**17%**
Self-satisfaction	**14%**	7%

*p<0.05 statistically significant

Figure 1 illustrates the frequencies of the merged *Motivation-Expectation* items. Since we found that most students cited "Knowledge enrichment" in both super-categories (*Motivation*- 93% of Graduates and 97% of Drop-outs; *Expectation*- 91% of Graduates and 93% of Drop-outs' question), we decided to remove this item from the graphical illustration (Figure 1), thus focusing on less frequent items. Besides the common items associated with *Motivation-Expectation*, only two additional items were included in the *Motivation* super-category: *Fondness of the domain* and *Creativity*, each of which also appeared in the *Aptitudes* super-category. Importantly, these items were mentioned in a slightly different context in both super-categories. For example, *Fondness of the domain* was mentioned as a motivating factor in the sense that when a student likes computer science, he may be motivated to develop a project (*Motivation*- 20% of the Graduates; 27% of the Drop-outs); on the other hand, *Fondness of the domain* was recognized as a significant factor to ensure success in the development process (*Aptitudes*- 30% of the Graduates and only 13% of the Drop-outs). The *Creativity* item was cited in the *Motivation* super-category in the context of: (a) the desire to express students' existing creativity skills; and (b) the desire to further develop those skills; both are difficult to achieve in the traditional school setting. This was more frequent when it was associated with *Aptitudes,* and it was slightly higher in the group of Graduates (25% of the Graduates; 20% of the Drop-outs).

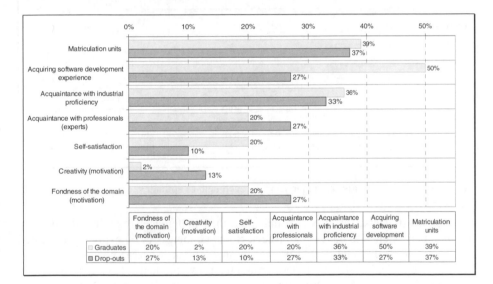

	Fondness of the domain (motivation)	Creativity (motivation)	Self-satisfaction	Acquaintance with professionals	Acquaintance with industrial proficiency	Acquiring software development	Matriculation units
Graduates	20%	2%	20%	20%	36%	50%	39%
Drop-outs	27%	13%	10%	27%	33%	27%	37%

Fig. 1. Frequencies of motivation and expectations (logically merged)

About a quarter of the students in both groups emphasized the importance of gaining matriculation units, thus revealing interest in the project development activity to achieve a practical outcome. Since additional matriculation units can promote students' possibilities for additional higher academic studies in many domains, we expected that a larger percentage of students would mention this factor. The relatively low percentage of the *matriculation units* item in comparison with other mentioned items implies that students' intrinsic motivation [13] was related to gaining knowledge and becoming acquainted with the "real" professional world of computing, which is perceived as more important (a similar percentage was found in both groups). Significant differences between the groups were noted regarding *Acquiring software development experience* (43% for Graduates; 20% for Drop-outs) and *Self-satisfaction* (14% for Graduates; 7% for Drop-outs) in the *Expectations* super-category. These items may be considered as possible triggers for the Graduates to complete their projects.

Figure 2 illustrates students' self-assessment of their aptitudes. Significant difference between the groups relates to *Fulfilling requirements* and *Fulfilling standards* aptitudes. Surprisingly, Drop-out students who apparently were unable to fulfill requirements needed to complete their projects, were more frequently stated, compared with Graduates, that they possess these aptitudes (*Fulfilling requirements*- 0% for Graduates, 13% for Drop-outs; *Fulfilling standards*- 0% for Graduates, 10% for Drop-outs).

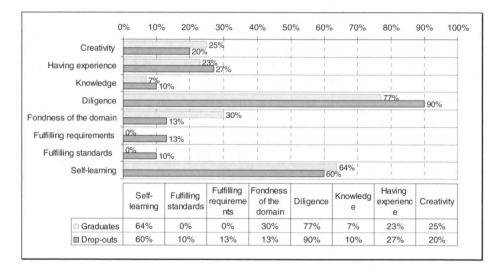

Fig. 2. Frequencies for each item of aptitudes

Sixty-two students (out of N=75) indicated their preferences regarding the domain in which they wanted to develop a project. Some Graduates (4%) explicitly indicated that they were ready to develop a project in any domain, whereas none of the Drop-outs indicated this. Besides, more Drop-out students were incapable of specifying any domain of interest (17% of Drop-outs; 9% of Graduates). We retrospectively examined the relationship between the student's initial domain of interest and the domain in which he actually developed her/his project. We assessed the strength of linkage between domains in terms of the following ranks: High, Medium, and Meaningless. We found that 35% of Graduates (out of N=45), compared with 26% of Drop-outs (out of N=30), developed a project in a domain with High strength of linkage to their initial domain of interest. When referring to the whole group of students who developed a project in a domain with High strength of linkage to their initial domain of interest (N=24), the ratio between Graduates and Drop-outs in that group is significantly high (2:1- 66% of Graduates; 33% of Drop-outs). Interestingly, the same ratio was found between Graduates and Drop-outs within the group of students who developed a project in a domain with Meaningless strength of linkage to their initial domain of interest (N=18).

Regarding pre-assessment of time needed for the project, we found that the groups differed (approaching statistical significant difference, t-test, p=0.06; Graduates: N=43 AVR=4.635 SD=2.54; Drop-outs: N=30 AVR=3.60 SD=1.91). The findings indicated that the graduates had a more realistic appreciation of the time they will need to invest in their project. The frequencies plot (Figure 3) illustrates that almost 40% of the Drop-outs group assessed the required investment in the project to be around one or two hours only. This may be one possible explanation why they gave up the project.

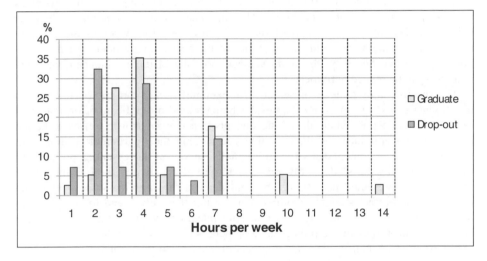

Fig. 3. Pre-assessment of time needed for the project

We can conclude that graduate students' initial motivation to develop a project was higher than the motivation of drop-out students. Graduates demonstrated a more open-minded approach and exhibited perspectives of professional experience compared with drop-out students. Specifically, Graduates presented a more flexible attitude towards the domain of interest in which they preferred to develop a project; they also had a more realistic appreciation of their capacity and of the time they will need to invest in their project.

It seems that one possible cause for the drop-out phenomenon may be the gap between Drop-outs' low appreciation of the requirements and, in contrast, their relatively high self-estimation of their aptitudes to fulfill requirements. We also found that almost half of the Graduates (43%) were aware of the importance of acquiring experience in software development (as opposed to only 20% of the Drop-outs). This may imply that graduate students have a greater awareness of one essential benefit accrued from participating in the project, namely, professional experience, and there is some evidence that they have some perspectives regarding the computing domain in their future.

4 Concluding Remarks

We presented an extracurricular program based on a combination of enrichment meetings and personal project development. The program aims at exposing talented high-school students, directly by leading experts, to state-of-the-art computing research, advanced technologies, software engineering methodologies, and professional norms. One main drawback in the stage of project development is the high number of drop-outs.

We believe that in order to ensure that the students will benefit as much as possible from the program, it is important to cultivate collaboration and establish good communication between all the "players" (students, mentors, teachers, and the leading team of the program). This can be fostered by accompanying the development of the program with a long-term formative evaluation while monitoring different processes that take place throughout its operation. The uniqueness of our evaluation model is its being naturally integrated as part of the program's operation, whereas the research tools serve as continuous "built-in" activities of the operational model. When evaluation is performed this way, the study subjects do not feel inconvenienced, and may benefit from the related activities.

The specific study that we presented here exemplifies such an approach since the questionnaire posed to the students, besides being a kind of application form, as well as a tool for collecting data, induced the students to perform intrinsic self-examination regarding their aptitudes and their readiness to commit their participation in a long-term project development activity, and actually served as an essential preliminary stage in preparing the students towards this activity. Importantly, the findings of the specific study presented here should be considered as a preliminary view regarding the factors that affect students' readiness to commit themselves and to complete activities of this kind. In order to obtain a more integrated view of the causes for the high rate of drop-outs in the project development activity, we plan to further qualitatively examine students' project development processes and their outcomes.

References

1. ACM/IEEE Joint Task Force on Computing Curricula, Software Engineering 2004 Curriculum Guidelines for Undergraduate Degree Programs in Software Engineering, A Volume of the Computing Curricula Series (August 2004)
2. The Ministry of Education, Israel, A high-school Software Engineering program (2004), http://csit.org.il (in Hebrew)
3. Bracken, B.: Progressing from student to professional: the importance and challenges of teaching software engineering. JCSC 19(2), 358–368 (2003)
4. Fincher, S., Petre, M., Clark, M. (eds.): Computer Science Project Work Principles and Pragmatics. Springer, London (2001)
5. Gal-Ezer, J., Beeri, C., Harel, D., Yehudai, A.: A high-school program in computer science. Computer 28(10), 73–80 (1995)
6. Gal-Ezer, J., Zeldes, A.: Teaching software designing skills. Computer Science Education 10(1), 25–38 (2000)
7. Haberman, B., Yehezkel, C.: Computer science, academia & industry – An educational program for establishing an entry point to the computing community of practice. J. Infomation Technology Education 7, 81–100 (2008)

8. Haberman, B., Yehezkel, C., Salzer, H.: Making the computing professional domain more attractive: an outreach program for prospective students. Int. Journal of Engineering Education 25(3), 534–546 (2009)
9. Haberman, B., Yehezkel, C.: Long-Term Software Projects Development – The Affect of Students' Self-Appreciation and Initial Expectations. In: ITiCSE 2009, Paris, France, p. 363 (2009)
10. Holcombe, M., Stratton, A., Fincher, S., Griffiths, G. (eds.): Projects in the computing curriculum. Proceedings of the Project 1998 Workshop. Springer, London (1998)
11. Long, P.D., Ehrmann, S.C.: Future of the learning space: Breaking out of the box. Educause, 42–58 (2005)
12. Passig, D.: Taxonomy of IT future thinking skills. In: Tailor, H., Hogenbirk, P. (eds.) Information and Communication Technologies in Education: The School of the Future, pp. 152–166. Kluwer Academic Publishers, Boston (2001)
13. Pollack, S., Scherz, Z.: Supporting project development in CS – the effect on intrinsic and extrinsic motivation. In: Proceedings of the 10th PEG Conference, Tampere, Finland, pp. 143–148 (2005)
14. Scherz, Z., Pollack, S.: An organizer for project-based learning and instruction in computer science. In: ITiCSE 1999 Conference, Cracow, Poland, pp. 88–90 (1999)
15. Sims-Knight, J.E., Upchurch, R.L.: Teaching software design: a new approach to high school computer science. In: Annual Meeting of the American Education Research Association, Atlanta, GA (April 1993), http://www2.umassd.edu/CISW3/people/faculty/rupchurch/papers/aera.pdf (last retrieved 26/10/09)
16. Yehezkel, C., Haberman, B.: Bridging the gap between school computing and the "Real world". In: Mittermeir, R.T. (ed.) ISSEP 2006. LNCS, vol. 4226, pp. 38–47. Springer, Heidelberg (2006)

Author Index

Printing: Mercedes-Druck, Berlin
Binding: Stein+Lehmann, Berlin